Geomorphology and the Carbon Cycle

RGS-IBG Book Series

For further information about the series and a full list of published and forthcoming titles please visit www.rgsbookseries.com

Published

Geomorphology and the Carbon Cycle

Martin Evans

WILEY

Registered Office
John Wiley & Sons, Inc., 111 River Street, Hoboken, NJ 07030, USA
John Wiley & Sons Ltd, The Atrium, Southern Gate, Chichester, West Sussex, PO19 8SQ, UK

Editorial Office
9600 Garsington Road, Oxford, OX4 2DQ, UK

For details of our global editorial offices, customer services, and more information about Wiley products visit us at www.wiley.com.

Wiley also publishes its books in a variety of electronic formats and by print-on-demand. Some content that appears in standard print versions of this book may not be available in other formats.

Library of Congress Cataloging-in-Publication Data
Names: Evans, Martin, 1970- author.
Title: Geomorphology and the carbon cycle / Martin Evans.
Description: Hoboken, NJ : John Wiley & Sons, 2022. | Series: RGS-IBG book series | Includes bibliographical references and index. | Summary: "As global atmospheric carbon concentrations continue to rise, there has been an increasing focus in the 21st century on understanding terrestrial components of the carbon cycle. This has been a major interdisciplinary research agenda and advances in remote sensing and modelling of vegetation systems have developed increasingly detailed understanding of above ground carbon cycling (Fatichi et al. 2019; Lees et al. 2018). Similarly, the storage of carbon in soils below ground has been the focus of extensive and detailed research (Wiesmeier et al. 2019). However, arguably understanding of soil carbon processes lags behind analysis of above ground systems. For example, it is notable that, in the paper cited at the top of this chapter (Bloom et al. 2016), the terrestrial carbon model that the paper applies includes significant detail around the cycling of carbon through biomass, modelling carbon in leaves, roots and wood separately, whilst soil carbon represents a single store. Where more detailed models of soil carbon cycling are applied that consider multiple solid carbon pools (e.g. Abramoff et al., 2018), a notable absence is consideration of lateral transfers of organic carbon in the soil and sediment system. Over the last ten years however, there has been an increasing recognition of the importance of lateral carbon fluxes within the landscape as a key part of understanding carbon dynamics at the large scale (e.g. Battin et al 2008). Figure 1.1 is the 5th Intergovernmental Panel on Climate Change (IPCC) representation of the terrestrial carbon cycle (IPCC 2013). Flux from the land to the oceans is represented by the fluvial carbon flux. Whilst the IPCC estimates distinguish pre-industrial and post-industrial fluxes for many of the key elements of the cycle, human impacts are not quantified for the fluvial system. Clearly, a more detailed picture of the fluvial system is required. The fluvial carbon flux is relatively small compared to the magnitude of terrestrial carbon storage, but is simply the residual of carbon transformation which occurs as organic matter is transported from headwaters to the oceanic sink. Much of the uncertainty about the relative importance of lateral carbon fluxes in the terrestrial carbon budget stems from a lack of knowledge about how large this residual is as a proportion of the total amount of organic carbon which is transported and delivered from hillslopes"— Provided by publisher.
Identifiers: LCCN 2021028996 (print) | LCCN 2021028997 (ebook) | ISBN 9781119393214 (hardback) | ISBN 9781119393252 (paperback) | ISBN 9781119393283 (pdf) | ISBN 9781119393245 (epub) | ISBN 9781119393290 (ebook)
Subjects: LCSH: Carbon cycle (Biogeochemistry) | Geomorphology.
Classification: LCC QH344 .E83 2022 (print) | LCC QH344 (ebook) | DDC 577/.144--dc23
LC record available at https://lccn.loc.gov/2021028996
LC ebook record available at https://lccn.loc.gov/2021028997

Cover image: Courtesy of Martin Evans
Cover design by Wiley

Set in 10/12pt PlantinStd by Integra Software Services Pvt. Ltd, Pondicherry, India

The information, practices and views in this book are those of the author and do not necessarily reflect the opinion of the Royal Geographical Society (with IBG).

SKY61EA52F8-40F3-4945-8003-E966C905DACF_012622

Contents

Series Editors' Preface

The RGS-IBG Book Series only publishes work of the highest international standing. Its emphasis is on distinctive new developments in human and physical geography, although it is also open to contributions from cognate disciplines whose interests overlap with those of geographers. The Series places strong emphasis on theoretically-informed and empirically-strong texts. Reflecting the vibrant and diverse theoretical and empirical agendas that characterize the contemporary discipline, contributions are expected to inform, challenge and stimulate the reader. Overall, the RGS-IBG Book Series seeks to promote scholarly publications that leave an intellectual mark and change the way readers think about particular issues, methods or theories.

For details on how to submit a proposal please visit: www.rgsbookseries.com

Ruth Craggs, *King's College London, UK*
Chih Yuan Woon, *National University of Singapore*
RGS-IBG Book Series Editors

David Featherstone
University of Glasgow, UK
RGS-IBG Book Series Editor (2015–2019)

Acknowledgements

This book has forced me to engage with a broad sweep of geomorphology. As a geographically trained geomorphologist I have been grateful, as I have carried out the research for this book, for the breadth of that education. I would like to acknowledge the influence of a number of remarkable geographers and geomorphologists (and one soil scientist!) that I have studied with, including Richard Crabtree, Tim Burt, June Ryder, Mike Church, Les Lavkulich and Olav Slaymaker. I think that their influence on my thinking is written through this book.

The ideas in this book have also been influenced by conversations in offices, seminar rooms, pubs, coffee shops and in the field with many of my colleagues over the years. I would particularly like to acknowledge the influence in various ways of Yvonne Martin, Steve Rice, Jeff Warburton, Fred Worrall and Tim Allott.

During the period of writing this book I have been involved in two very stimulating workshop series. Colleagues who contributed to the MadCaP seminars in Manchester and to the NERC supported Peatland Resilience and Microbial Processes workshops have helped to develop my thinking around the key importance of understanding the interrelation of geomorphological and microbial processes as drivers for terrestrial carbon cycling.

One of the joys of an academic career is the opportunity to work with and to learn from brilliant graduate students. I have been very fortunate in this regard and I would like to thank former and current students (several of whom are now colleagues) Juan Yang, James Rothwell, Amer Al-Roichdi, Alan Clarke, Laura Liddaman, Eleanor Teague, Sarah Crowe, Steve Daniels, Richard Pawson, Claire Goulsbra, Emma Shuttleworth, Beth Lowe, Andrew Stimson, Donald Edokpa, Sarah Brown, Jane Mellor, Dylan Zhang, Adam Johnson and Richard Figuera for myriad conversations from which I have learnt an enormous amount.

The figures for this volume have been prepared by Nick Scarle and Graham Bowden in the cartographic unit in the department of geography in Manchester.

Their skill and attention to detail in bringing the illustrations to life is very much appreciated.

My research has been supported in many ways over the years by colleagues in the geography laboratories in Manchester whose expertise in the lab and in the field is invaluable. John Moore, Martin Kay, Jonathan Yarwood and Tom Bishop in particular.

My research has been funded by a wide range of bodies including NERC, DEFRA, Environment Agency, United Utilities, Natural England, Moors for the Future, the Royal Society, Manchester University, The British Society for Geomorphology, The Royal Geographical Society and the Leverhulme Trust, and I am very grateful for the support they have provided to explore some of the challenges outlined in this book. I am particularly grateful to the Leverhulme Trust who granted me a one-year fellowship during which a significant amount of the work on this volume was completed.

The most significant acknowledgement for this book is to Danielle Alderson as a PhD student and as a colleague. Discussions with Danielle have significantly shaped my thinking on geomorphology and carbon cycling, particularly in fluvial contexts. She has copy-edited all of the chapters, made useful comments, and helped with many of the practicalities of preparing the manuscript. I am enormously grateful to her for her contribution.

I would also like to thank Bob Hilton and an anonymous reviewer for a really helpful set of comments on the first draft of the manuscript which has significantly improved the final draft.

Finally, I would like to thank my family for putting up with me and 'the book' for so long!

Part I
The Terrestrial Carbon Cycle and Geomorphological Theory

Chapter One
Geomorphology and the Terrestrial Carbon Cycle

The terrestrial carbon cycle is currently the least constrained component of the global carbon budget.

(Bloom et al. 2016: 1285)

Introduction

As global atmospheric carbon concentrations continue to rise, there has been an increasing focus in the twenty-first century on understanding terrestrial components of the carbon cycle. This has been a major interdisciplinary research agenda and advances in remote sensing and modelling of vegetation systems have developed increasingly detailed understanding of above ground carbon cycling (Fatichi et al. 2019; Lees et al. 2018). Similarly, the storage of carbon in soils below ground has been the focus of extensive and detailed research (Wiesmeier et al. 2019). However, arguably, understanding of soil carbon processes lags behind analysis of above ground systems. For example, it is notable that in the paper cited at the top of this chapter (Bloom et al. 2016), the terrestrial carbon model that the paper considers includes significant detail around the cycling of carbon through biomass, modelling carbon in leaves, roots and wood separately, whilst soil carbon represents a single store.

Where more detailed models of soil carbon cycling are applied that consider multiple solid carbon pools (e.g., Abramoff et al. 2018), a notable absence is consideration of lateral transfers of organic carbon in the soil and sediment

Geomorphology and the Carbon Cycle, First Edition. Martin Evans.

system. Over the last ten years however, there has been an increasing recognition of the importance of lateral carbon fluxes within the landscape as a key part of understanding carbon dynamics at the large scale (e.g., Battin et al. 2008). Figure 1.1 is the 5th Intergovernmental Panel on Climate Change (IPCC) representation of the terrestrial carbon cycle (Cubasch et al. 2013). Flux from the land to the oceans is represented by the fluvial carbon flux. Whilst the IPCC estimates distinguish pre-industrial and post-industrial fluxes for many of the key elements of the cycle, human impacts are not quantified for the fluvial system. Clearly a more detailed picture of the fluvial system is required. The fluvial

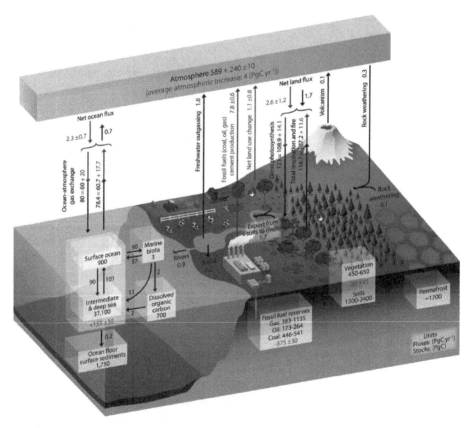

Figure 1.1   A simplified schematic of the global carbon cycle. Black text indicates pre-industrial stores and fluxes and grey indicates estimated changes post circa 1750. Source: After Ciais et al. 2013. Figure 6.1 in *Climate Change 2013: The Physical Science Basis. Contribution of Working Group I to the Fifth Assessment Report of the Intergovernmental Panel on Climate Change* (p. 471). Reproduced with permission of Cambridge University Press. (https://www.ipcc.ch/site/assets/uploads/2018/02/WG1AR5_Chapter06_FINAL.pdf)

carbon flux is relatively small compared to the magnitude of terrestrial carbon storage, but is simply the residual of carbon transformation which occurs as organic matter is transported from headwaters to the oceanic sink. Much of the uncertainty about the relative importance of lateral carbon fluxes in the terrestrial carbon budget stems from a lack of knowledge about how large this residual is as a proportion of the total amount of organic carbon which is transported and delivered from hillslopes.

Organic carbon in solid or particulate form is transported from hillslopes to the fluvial system, and can be transformed or mineralised in transit either physically, or through the action of macro and micro biota. Transit of organic matter across hillslope systems is however complex and variable in time and space. The proportion of organic sediment eroded in a given period or event that is delivered to the river system (the sediment delivery ratio Walling 1983) is considerably less than 100%, so that an understanding of hillslope geomorphology is required to determine where eroded carbon is deposited. Once organic matter reaches the river system, timescales for direct transfer of dissolved and suspended material to the ocean are typically hours to days (e.g., Jobson 2001). However, Ferguson (1981) described the fluvial sediment transport as a 'jerky conveyor belt' so that a proportion of sediment is redeposited within the fluvial system (on bars or floodplains, or in lakes or reservoirs). Material may be mobilised and redeposited multiple times before reaching the ocean, so that virtual velocities may drop by several orders of magnitude and travel times are consequently measured in centuries rather than days. The transit of organic carbon from hillslope source (involving the processes of carbon fixation by vegetation and transfer of litter to the soil system) to oceanic sink is complex. Along the way, material may be stored in zones of sediment accumulation (depositional landforms), representing long-term carbon sequestration or alternatively may be mineralised and lost to the atmosphere through processes of microbial decomposition and respiration. The interactions of carbon fixed by the terrestrial biosphere within the sedimentary system are significantly more complex than the representation in the IPCC carbon budget.

Initially work on lateral transfers of carbon tended to be focussed on the transformation of organic carbon within freshwater systems, building on concepts such as the River Continuum Concept (Vannote et al. 1980). This concept postulates predictable patterns of downstream change in organic matter quality as it is cycled by in-stream processes. Increasingly however, the role of geomorphological processes in controlling the lateral transfer of carbon on hillslopes and through river systems has been recognised (e.g., M. Evans et al. 2013; Hoffmann et al. 2013a). Agricultural hillslope systems have been a major focus of geomorphological work in this area and anthropogenic modification of these systems constitute a major alteration of the overall terrestrial carbon cycle. However, a focus on agricultural systems has tended to drive a focus on field scale patterns of sediment transfer.

Unpicking the black box of headwater to ocean carbon transfers implied in Figure 1.1 requires consideration of carbon fluxes across the entire sediment cascade. A rapidly expanding body of geomorphological research has begun to explore the role of the sediment system in the terrestrial carbon cycle (e.g., Hoffmann et al. 2009; Kirkels et al. 2014). However, the focus of geomorphological carbon cycling research has predominantly been on characterising the magnitude of carbon storage in major loci of sediment accumulation (depositional landforms). There is increasing recognition that a complete understanding of sedimentary carbon storage also requires analysis of rates of carbon addition and removal from storage, and the processes which control this. This requires an integration of biological and geomorphological analyses. Nevertheless, despite the call by Slaymaker and Spencer (1998) for biogeochemical cycling to become a central concern of physical geography and geomorphology, the wider engagement of geomorphology with an understanding of biogeochemical cycling has been so far limited.

In the context of a rapidly shortening time horizon for effective action to mitigate rising greenhouse gas concentrations in the atmosphere, it is argued that rapid progress in this area is vital. The requirement to deliver a more complete understanding of the terrestrial carbon cycle has two main components. Firstly, a functional understanding of the processes which drive carbon flux through the terrestrial system is needed in order to understand the interaction of the terrestrial biosphere with excess atmospheric carbon derived from fossil fuel use. In particular, a focus on carbon storage and release in the terrestrial system is fundamental to identifying positive feedback mechanisms and threshold conditions that might exacerbate anthropogenically driven rates of change. Secondly, understanding the processes by which carbon is added to major terrestrial carbon stores, such as live biomass and particularly soil and sediment carbon storage, offers the potential to manipulate these natural systems to sequester carbon and therefore potentially provide some mitigation of rising atmospheric carbon levels.

Increasing amounts of academic labour are being focussed on these critical problems, but arguably too much of this work is siloed within traditional disciplinary structures and networks. This book is written from the perspective of a geomorphologist trained in geography departments in the UK and Canada, and is born partly from the conviction that there is much which traditional geomorphological understanding of landscape systems can bring to the grand challenge of understanding and managing terrestrial carbon storage. A multidisciplinary approach is fundamental to meeting this challenge; this book will both explore and explain the ways in which geomorphological understanding can contribute to, and also identify the challenges of integrating this knowledge, with understanding of the biosphere and its role in the fixation and release of atmospheric carbon.

The Aims of This Book

The overall aim of this book is to develop a research agenda for the integration of geomorphological, biological and microbiological understanding into analyses of the terrestrial organic carbon cycle. To achieve this, this book has three main objectives:

1) To identify challenges and opportunities in the application of geomorphological methods and insights to the analysis of terrestrial carbon cycling.
2) To synthesise the rapidly expanding understanding of geomorphology and carbon cycling in the academic literature to define the state of the science.
3) To develop a conceptual framework based on geomorphological theory, and informed by work in ecology, microbiology and biogeochemistry, in order to analyse spatial patterns of terrestrial carbon cycling at the landscape scale.

Achieving the aim of fully integrating geomorphological expertise into a multidisciplinary approach to the analysis of terrestrial carbon cycling will not happen just because this integration offers answers to key questions, it also requires an understanding across scientific communities of what those questions are. This book is written with three audiences in mind. It is written for geomorphologists, to provide a synthesis for those in the field, but also to persuade the wider geomorphological community that core geomorphological data, skills and understanding are required to untangle the complexities of the terrestrial carbon cycle, and that they hold the key to progress in this area.

This book is also written for biologists and microbiologists whose work drives our understanding of both the fixation of atmospheric carbon and the mineralisation of soil carbon, which underpins carbon sequestration into sedimentary stores. This book will make the case to this audience that the physical movement of carbon substrate across the landscape, and the disturbance of equilibrium communities that are associated with phases of erosion and deposition, are components of carbon cycling which need to be integrated into an understanding of carbon storage at longer timescales. These decadal and longer timescales are critical in the context of anthropogenically driven changes in atmospheric carbon concentrations over similar periods.

Finally, this book aims to engage with the community who manage the terrestrial system. Landowners, planners and policy makers are the people who have the capacity to effect change in the anthropogenically dominated landscape that we live in. Managing erosion through landscape restoration, re-naturalising river courses and modifying agricultural practices are all components of conservation practice which drive change in biological and geomorphological systems, and thus modify the flux of carbon through terrestrial systems. Understanding these changes offers the potential to design such interventions in ways which maximise carbon sequestration into sedimentary stores, and so have potential to mitigate some anthropogenic carbon emissions. By characterising these stores and the processes which control carbon sequestration at a range of timescales, this book will offer the potential to argue for carbon sequestration co-benefits in landscape conservation schemes.

Organisation and Focus of This Book

The main argument of this book is developed in three parts. **Part I** (Chapters 2–4 of this book) outlines the key elements of the fast (organic) and slow (inorganic) terrestrial carbon cycles (Chapters 2 and 3 respectively), in order to provide the context for the discussion of geomorphological influences on carbon cycling in the subsequent chapters. The main focus in this book is on the fast carbon cycle, but a review of the key elements of the slow cycle is important context for understanding what follows and is also included for completeness since this is an area where geologically trained geomorphologists are driving key elements of the research agenda.

One of the assumptions behind this book is that the processes which drive the fluxes within and the reorganisation of sediment systems are a major influence on carbon cycling. If this is the case, then the developed techniques and gathered process-based knowledge of more than a century of geomorphological research will make a contribution to a fuller explanation of the terrestrial carbon cycle. Therefore, in Chapter 4 a range of key conceptual approaches which underpin modern geomorphological thinking are highlighted and the ways in which they can contribute to understanding of carbon cycling are explored.

Part II (Chapters 5 to 8 of this book) focusses in more detail on the fast carbon cycle and the ways in which geomorphological processes interact with vegetation and soil microbiota to cycle carbon through the terrestrial system. Two organising principles underpin the concept of the sediment cascade and the idea of a carbon landsystem.

The Sediment Cascade

In Figure 1.1, the fluvial system which is the primary conduit for the direct transfer of carbon from the continents to the oceans is represented by a simple line. This is in contrast to the detail on the land surface, which indicates a range of processes driving terrestrial carbon cycling. However, as briefly discussed above, the fluvial system is complex and dynamic, and the previous model of the fluvial system as a pipe, simply transporting carbon from hillslope to ocean cannot be sustained. One of the organisational principles of this book is the

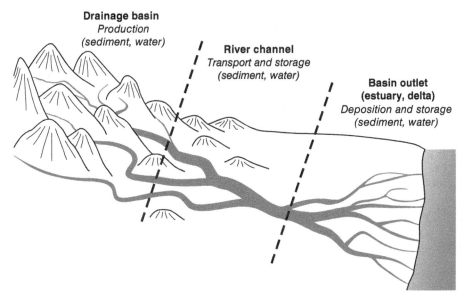

Figure 1.2   The sediment cascade. Source: After Schumm 1977. Reproduced with permission of The Blackburn Press.

sediment cascade (Burt and Allison 2010; Schumm 1977). Figure 1.2, from the paper by Schumm which initially outlined the concept, identifies key features of the cascade and defines a linear cascading system describing the flux of water and sediment through landscape systems, from production on hillslopes to deposition in oceans and estuaries. Figure 1.2 indicates that the key role of hillslopes and headwaters in the sediment system is the production of sediment, and the delivery of this material to downstream fluvial systems. In the context of organic carbon fluxes, the River Continuum Concept makes similar assumptions. However, as outlined in a recent analysis by Joyce et al. (2018), the flux of sediment through the upland sediment cascade involves both production of sediment through upland erosion, but also storage in a range of depositional landforms (Figure 1.3).

A full description of the flux of sediment and organic carbon through the landscape therefore requires analysis of all components of the sediment cascade, considering the production, storage and cycling of carbon. Chapters 5–8 review the literature on carbon cycling for four key components of the sediment cascade. These are hillslopes (Chapter 5), headwaters (Chapter 6), the fluvial system (Chapter 7) and estuaries/coasts (Chapter 8). For these specific contexts the chapters explore the interaction of geomorphological processes, vegetation and succession as a control on primary production and decomposition and mineralisation of organic carbon by the microbial system.

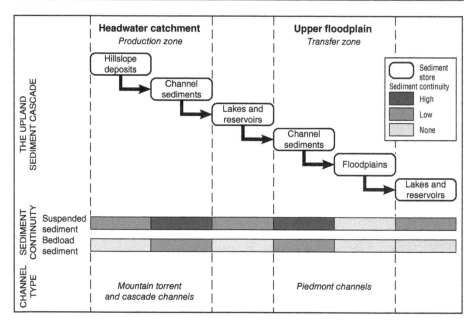

Figure 1.3   A model of the upland sediment indicating the complexity and continuity of sediment transfer in this sub-component of the wider cascade. Source: After Joyce et al. 2018 (https://doi.org/10.1016/j.geomorph.2018.05.002). Licensed by CCA 4.0.

The Carbon Landsystem

The review and synthesis of the literature in Chapters 5–8 aims to identify the primary carbon stores within each sub-component of the sediment cascade, and to analyse the key processes which drive translocation and transformation of organic carbon within the system. In the geomorphological literature, the term landsystem is used to describe conceptual models of synthetic landscapes that describe key landforms and the processes that drive material flux through the system and produce these characteristic landforms (e.g., Evans 2014). The analysis in this book builds from the assumption that the geomorphological context is not a boundary condition to the terrestrial carbon cycle, but is a dynamic driver of the flux of carbon through the sediment cascade, so that quantification of carbon flux requires integration of geomorphological, biological and microbiological processes. In this context, it is a short conceptual step to move beyond the landsystem as a description of characteristic geomorphologies to thinking about the carbon landsystem as the characteristic set of interactions between these three process types, which drive fluxes of carbon through depositional landforms at points along the sediment cascade. Chapters 5–8 attempt to characterise the carbon landsystem for key elements of the cascade.

Part III (Chapters 9 and 10 of this book) explores some of the implications of a geomorphological approach to understanding the carbon cycle from the perspective of the management of carbon landsystems, in order to mitigate anthropogenically driven increases in atmospheric carbon content. With the International Chronostratigraphic Commission looking likely to ratify a recommendation that the Anthropocene be recognised as a new geological epoch (Zalasiewicz et al. 2017), there is an emerging consensus that human action is the dominant control on environmental systems. As discussed above, one of the key drivers for developing an understanding of the functioning of carbon landsystems is to have the tools to actively manage these systems. Chapter 9 reviews progress in this direction. Finally, in Chapter 10 the benefits of integrating geomorphological understanding into our analysis of the terrestrial carbon cycle are summarised and conceptual and practical approaches to the concurrent analysis of biological, microbiological and geomorphological components of the carbon landsystem are proposed.

Chapter Two
Geomorphology and the Fast Carbon Cycle

Introduction

Carbon is cycled through the terrestrial system at a range of time and space scales via a wide variety of processes. A distinction is commonly made between the slow carbon cycle, which occurs over timescales of hundreds of thousands of years, cycling carbon between the lithosphere, the oceans and the atmosphere, through processes of weathering and sedimentation, and the fast carbon cycle, which operates at timescales of seconds to millennia, and involves the transfer of carbon between soils, vegetation and the atmosphere dominantly controlled by biological processes.

This chapter outlines the key processes which drive the fast carbon cycle and identifies major interactions between geomorphological processes and terrestrial carbon cycling. The fast carbon cycle is effectively the biological carbon cycle. Carbon is removed from the atmosphere photosynthetically by plants, both on land and in the oceans. It is also returned to the atmosphere through plant and animal respiration processes, which mineralise organic carbon and release carbon dioxide and methane to the atmosphere (Figure 2.1).

Photosynthesis

Organic carbon is fixed into living vegetation through photosynthesis. Photosynthesis is a critical component of the fast carbon cycle since it removes gaseous carbon from the atmospheric store (carbon dioxide) and fixes it into solid organic matter in plant tissues. Photosynthesis is a reduction reaction facilitated by chlorophyll

Geomorphology and the Carbon Cycle, First Edition. Martin Evans.

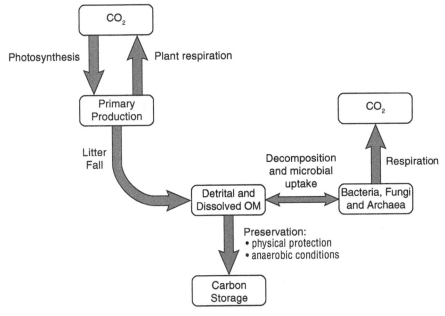

Figure 2.1   The fast carbon cycle. Source: Martin Evans.

in green plants. Chlorophyll is a pigment which absorbs light energy to utilise in enzyme mediated reactions, which fix carbon into carbohydrate and releases oxygen, as shown in Equation 2.1:

Equation 2.1: $6CO_2 + 6H_2O + \text{photons} \rightarrow C_6H_{12}O6 + 6O_2$

The primary environmental controls on the photosynthetic reaction are light intensity, carbon dioxide concentration and temperature. Water supply (soil moisture) is also important since it controls the opening and closure of leaf stomata, through which leaves exchange the gaseous photosynthetic reactants and products (CO_2 and O_2).

The critical control of the biosphere and photosynthesis over the fast carbon cycle is apparent from monthly atmospheric CO_2 data (Figure 2.2). The seasonal cycle of CO_2 flux is driven by enhanced photosynthetic fixation of carbon and lower atmospheric carbon dioxide concentrations during the northern hemisphere summer (which has a greater impact due to the hemispheric distribution of land masses).

Globally, Beer et al. (2010) estimate that total fixation of carbon by terrestrial plants is 123 ± 8 PgC per year, with 40% of this uptake occurring in tropical forests. This Gross Primary Productivity (GPP) figure does not represent the net carbon flux from the atmosphere because it does not take account of carbon release from vegetation by autotrophic respiration. The balance of these two fluxes is referred to as Net Primary Productivity (NPP).

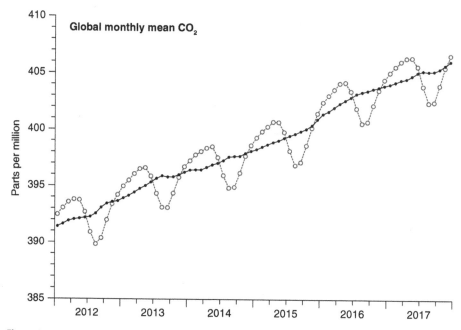

Figure 2.2   Mean global CO_2 concentration 2012–2017. Superimposed on the rising trend is a clear seasonal pattern. Source: After NOAA (https://www.esrl.noaa.gov/gmd/ccgg/trends) Public Domain Data.

Estimates of global terrestrial NPP are derived from a range of modelling approaches. Models typically are driven by light intensity (usually measured as Photosynthetically Active Radiation (PAR) which is defined as wavelengths between 400–700 nm). The ability of plants to photosynthesise carbohydrates from this light energy is controlled by plant physiology. Climatic variables, particularly temperature and precipitation, are often taken as proxies for plant growth in simple models, although more complex approaches have process-based sub-models describing plant function, and allocate material between soils, the atmosphere and vegetation (Cramer et al. 1999). In a major model comparison study, Cramer et al. (1999) ran 17 models spanning 3 main approaches (satellite-based, fixed plant structure models and dynamic vegetation models) (Figures 2.3a and 2.3b). Model outputs predict a global total terrestrial NPP of 44.4–66.3 PgC per year. This is approximately 10% of the IPCC estimate of global carbon storage in vegetation (Cubasch et al. 2013). It is a characteristic of the global carbon cycle that net gaseous carbon fluxes are small residuals from the difference between carbon uptake and release. As both of these fluxes are large, the sensitivity of this residual to proportionally small changes in uptake or release may be substantial. In the oceans, primary productivity is of a similar magnitude to the terrestrial system, but lower rates of NPP are balanced by the larger area of the ocean. Estimates range from 38–65 PgC yr^{-1} (Buitenhuis et al. 2013).

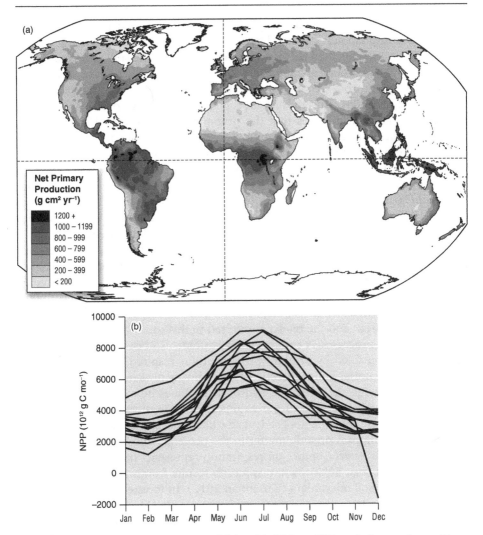

Figure 2.3   (a) Global photosynthesis from 17 global models; (b) Annual NPP g cm² a⁻¹ averaged across 17 models. Source: After Cramer et al. 1999. Reproduced with permission of Wiley.

Heterotrophic, Soil and Ecosystem Respiration

Ecosystem respiration is the sum of autotrophic respiration in plants and heterotrophic respiration by decomposer organisms. Net Primary Productivity (NPP) is the balance of Gross Primary Productivity (GPP equivalent to total photosynthetic fixation) and autotrophic respiration. This means that net carbon balance can be conceptualized either as the balance of GPP and ecosystem respiration or as the balance of NPP and heterotrophic respiration. Another term which is widely used is 'soil respiration'. Soil respiration is measured by release of CO_2 to the atmosphere at the soil surface. This represents total below ground respiration or the sum of heterotrophic respiration and plant root respiration (cellular respiration in the roots of vegetation).

Globally, plant root respiration is 30–50% total soil respiration (Bond-Lamberty et al. 2004). Recent data on the ratio of heterotrophic respiration to total soil respiration indicates that the proportion of heterotrophic respiration in soil respiration is increasing over time in response to climate warming with increases from 54% to 63% between 1990 and 2014 (Bond-Lamberty et al. 2018). This represents an increasing flux of carbon to the atmosphere through heterotrophic respiration of soil carbon. Global estimates of soil respiration are 68–98 PgC per year (Jian et al. 2018). Global autotrophic respiration is estimated at 64 ± 12 PgC per year (Ito 2020). The IPCC estimate total respiration and biomass burning at 118.7 PgC per year (Cubasch et al. 2013 and Figure 1.1).

Accumulation of carbon in the biosphere through NPP is offset by gaseous carbon release via mineralisation of soil carbon through heterotrophic respiration of soil detritivores. Decomposition and mineralisation of detrital organic matter is dominated by heterotrophic microbes (primarily bacteria and fungi). An excellent summary of microbial processes controlling soil respiration is given by Kirchman (2012) and the reader is referred to this source for a detailed introduction to these processes. A summary of key points is presented below.

Organic matter fixed by primary productivity is respired by microbial biota. Large detrital material is not directly utilised, but is broken down into smaller components through the release of hydrolase enzymes by microbes (Burns and Dick 2002). Organic matter in the size range < 500 Da is transported across the cellular membrane and metabolised inside bacterial and fungal cells. Aerobic respiration decomposes organic matter in the presence of oxygen releasing CO_2 and H_2O. Not all heterotrophic soil respiration represents the decomposition of old carbon, as at least 50% of soil respiration results from the metabolisation of root exudates by soil microbiota (Kirchman 2012). In forested systems, Högberg et al. (2001) have demonstrated that cutting the supply of photosynthates to root systems by removing tree bark results in a 54% reduction of soil respiration. This means that modern root exudates are the dominant substrate for microbial respiration (particularly by mycorrhizal fungi). Consequently, much of the carbon which is respired back to the atmosphere is rapidly turned over from the soil (through mineralisation of modern labile carbon from the root exudates) to the atmosphere. For these reasons, NPP and respiration are strongly linked at both the ecosystem and global scale (Figures 2.4a–2.4c).

In aqueous systems, bacteria dominate microbial decomposition of organic matter because they are better able to compete for small dissolved molecules. In contrast, in terrestrial systems fungi may represent over half the microbial biomass, because fungal hyphae have the ability to span dry microsites in the soil system, and access remote sources of metabolites (Kirchman 2012). Fungi are also able to

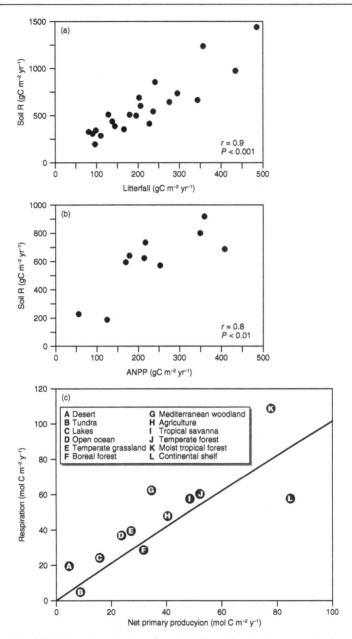

Figure 2.4   (a) and (b) Relation between above ground NPP/litterfall and soil respiration for forests and grasslands. Source: After Raich and Tufekciogul 2000. Reproduced with permission of Springer. (c) NPP vs. respiration across a range of global biomes. Source: After Kirchman 2012. Reproduced with permission of Oxford University Press.

degrade recalcitrant materials in terrestrial environments such as lignin, whereas bacteria respire simpler and more labile organic compounds. However, research in arctic soils has indicated that fungi are also able to utilise available simple compounds (Rinnan and Bååth 2009). In addition, fungi are also more drought tolerant than bacteria (de Vries et al. 2012, 2018; Yuste et al. 2011), which means that they are more able to tolerate fluctuating moisture conditions and potentially play an important role in cycling carbon in disturbed systems.

Decomposition of organic matter by aerobic respiration in the presence of oxygen is an oxidation reaction where carbon is oxidised to CO_2. This redox reaction generates electrons which are moved through a biochemical electron transport chain to a final electron acceptor. In the process of aerobic respiration, oxygen is the electron receptor. Under anaerobic conditions, carbon dioxide can still be produced from carbon and oxygen atoms within the organic matter, but an alternative electron acceptor is required. Nitrates and sulphates are common ionic species that perform the role of an electron acceptor. Anaerobic microorganisms (particularly bacteria and Archaea) respire in low oxygen environments using these alternative electron acceptors.

In the context of the carbon cycle, the most important mechanism of anaerobic respiration is production of methane by methanogenic microorganisms (primarily Archaea). Methanogenesis occurs in very low redox conditions once all molecular oxygen, nitrate, iron, manganese and sulphur has been reduced (Smith et al. 2003). Anaerobic decomposition of organic matter is very slow and so saturated reducing environments such as deep peats preserve organic matter and store organic carbon fixed from the atmosphere. For example, Lee et al. (2012) report rates of carbon release from thawed permafrost which were 4–10 times more rapid under aerobic conditions. Even correcting for the fact that aerobic emissions are predominantly CO_2 rather than CH_4, greenhouse warming potential from melted permafrost in aerobic conditions had a climate forcing 1.5–7.1 times higher than in anaerobic conditions.

Despite the rapid decomposition of organic matter under aerobic conditions, the high greenhouse gas potential of methane results in the process of methanogenesis of organic matter in reducing conditions, having the potential to produce substantial impacts on atmospheric greenhouse gas budgets. However, as a consequence of oxidation, methane has a shorter residence time in the atmosphere than CO_2 (about 12 years: Myhre et al. 2013), so its relative influence on greenhouse forcing is reduced at longer timescales. For example, considering the impact of CO_2 sequestration and CH_4 production in peatlands, Whiting and Chanton (2001) concluded that at 20 year timescales methane production increased greenhouse warming potential, whereas over 100 years, temperate and subtropical peats provide a net cooling influence on the atmosphere, and northern wetlands are greenhouse neutral. At 500-year timescales, all peatlands were found to have a net climate cooling effect. In this study, warmer temperatures in the wetlands with longer summer growing seasons fixed more CO_2 and sequestered more carbon, hence their greater cooling potential.

Controls on Heterotrophic Respiration of Soil Carbon

Soils are complex biogeochemical systems, so that the controls on rates of soil respiration are numerous and vary substantially in both time and space. For example, Bragazza et al. (2013) describe complex feedbacks between warming temperatures, increasing prevalence of Ericaceous shrubs on peatlands and changes in microbial populations, which favour enhanced decomposition of soil organic matter. Similarly, Van Der Heijden et al. (2008) argue that soil microbial populations are key drivers of plant productivity through their role in facilitating and competing for nutrient supplies (nitrogen and phosphorus).

Vegetation is an important control on rates of soil respiration, because of the linkages between plant and microbial populations, and also because of the key role of root exudates as substrates for microbial respiration. However, it has been argued that at the global scale the influence of vegetation type on respiration rates is secondary to moisture and temperature controls (Raich and Tufekciogul 2000).

Temperature is a primary control on rates of soil respiration (Davidson and Janssens 2006; Lloyd and Taylor 1994; Rustad et al. 2000). Respiration is a chemical process biologically mediated by enzyme activity, so that the temperature dependence of chemical reactions and enzyme activity drive this relation. There is a strong positive correlation between mean annual air temperature for global biomes and reported rates of soil respiration (Raich and Schlesinger 1992).

The temperature dependence of respiration is commonly reported as Q10 values, where these represent the proportional increase in respiration for a 10 degree rise in temperatures (Figure 2.5b). Kirschbaum (1995) reviewed global Q10 data and showed that values decline non-linearly from values of eight at temperatures of 0°C to less than two at 35°C. This non-linearity is potentially important as a control on changing spatial patterns of soil respiration in response to global change, since rapid warming of high latitudes (Lloyd and Taylor 1994; Pithan and Mauritsen 2014) is coincident with climate regimes where higher Q10 values are common, and additionally with extensive storage of soil carbon across the northern peatlands (Gorham 1991).

Conant et al. (2011) identify three key processes which mediate the temperature response: 1) rates of enzymatic depolymerisation of large molecules; 2) rates of microbial enzyme production; and 3) processes which limit availability of soil organic matter to microbial action, including adsorption to mineral surfaces and soil aggregate turnover (summarised in Figure 2.6). However, there is not a simple relation between temperature and respiration because of other environmental influences on the rate of reaction. Kirschbaum (2006) emphasises the importance of moisture regime and substrate availability in modifying the relation, whilst Conant et al. (2011) have modelled variable temperature dependence of soil respiration based on characterisation of either physical or chemical mechanisms protecting soil organic matter from decomposition, as described below.

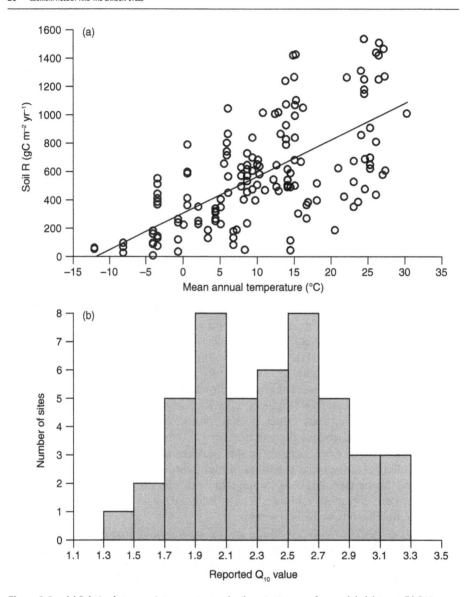

Figure 2.5 (a) Relation between air temperature and soil respiration rates from a global dataset; (b) Q10 values from a global dataset of soil respiration rates. Source: After Raich and Schlesinger 1992. (https://doi.org/10.3402/tellusb.v44i2.15428) CC-BY.

After temperature, moisture is the best studied biophysical control on soil respiration rates; arguably in terms of local spatial variability in respiration rates, it is the most dominant. The principle mechanism underlying moisture control on respiration is through control on oxygen concentrations in saturated soils (since

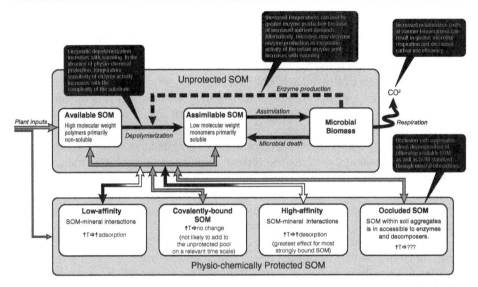

Figure 2.6 Controls on soil organic matter decomposition. Source: After Conant et al. 2011. Reproduced with permission of Wiley.

respiration requires an electron acceptor). Local accumulation of soil organic matter in wetland soils occurs because saturation limits rates of microbial decomposition. In peatlands with saturated soils, modelling of soil respiration rates is improved over simple temperature dependent approaches (Lloyd and Taylor 1994) through the addition of a water table term (Rowson et al. 2013). The response of respiration rate to moisture is however not simply bimodal between saturated and unsaturated conditions. Linn and Doran (1984) demonstrated that maximum activity by aerobic microbes occurs at around 60% pore water saturation, with activity at higher saturations limited by reduced oxygen content, and at lower saturations by water content associated limitations on microbial activity.

The combined importance of temperature and moisture as controls on respiration rates have also been explored in the context of a topographic component. Altitudinal variations in temperature, and the control of slope angle, slope form and slope position on soil moisture (Burt and Butcher 1985) mean that there is potentially strong topographic control on spatial patterns of respiration rate. This has been demonstrated at the hillslope scale in forests in Korea by Kang et al. (2003), who conclude that Q10 based models of soil respiration rates are inadequate for prediction at large spatial scales, unless topographic controls on soil moisture are considered.

Microbial Population Controls on Heterotrophic Soil Respiration

Wieder et al. (2013) argue that conventional modelling approaches to turnover of soil carbon, which are primarily driven by the nature of the organic substrate and simple kinetic models of decomposition rate, are insufficient to capture the response of heterotrophic soil respiration to changing external environments. This is because the models do not capture microbial physiological responses to those changes. Conventional wisdom suggests that microbial responses to changes in environment are rapid, so that models typically assume that microbial processes respond almost instantaneously to changes in the biophysical environment. However, recent work (e.g., Hawkes et al. 2017) suggests that lags in soil respiration response to changes in moisture might relate to much slower changes in microbial populations, which show resilience to environmental change at timescales of over a year. In this experimental work, respiration in grassland sites was controlled in part by historic climate, with the response of respiration to moisture changes shifting four-fold across a 480 mm rainfall gradient, so that wet soils have lost twice the carbon of historically dry soils. This example emphasises the complexity of soil respiration responses to environmental change, with microbial population dynamics and autecology potentially playing a key role alongside biochemical responses to the changing physical environment. Hawkes et al. (2017) suggest that such lags in microbial population response are one explanation of the observation that even models of soil carbon stocks that explicitly consider microbial processes, are only able to explain around half of observed spatial variation (Wieder et al. 2013).

Interactions of Environmental Controls on Heterotrophic Soil Respiration

On a global scale, spatial variability in temperature and moisture, and presumably of microbial populations (although this has not been systematically reviewed) control heterotrophic soil respiration rates, so that rates of soil carbon turnover vary from 500 years in tundra and peatlands to just 10 years in tropical savannah (Raich and Schlesinger 1992). Similarly, Carvalhais et al. (2014) demonstrate spatial variation in rates of soil carbon turnover from values of circa 15 years in the tropics, and up to 225 years at high latitudes.

Lower rates of respiration in colder and wetter environments mean that under these circumstances, net primary productivity can exceed heterotrophic respiration, so that carbon accumulates in soil profiles. However, because productivity and respiration are typically correlated, small changes brought about by inter-annual climatic variability can lead to variability in the rate of carbon sequestration or loss. Even peatlands which are typically sites of long-term carbon accumulation can become net sources of carbon to the atmosphere in some years (Worrall et al. 2009).

An influential review of controls on rates of soil organic matter decomposition by Davidson and Janssens (2006) frames the relation between the temperature dependence of decomposition and other environmental drivers, in the context of environmental controls on temperature sensitivity. It identifies five key environmental constraints on decomposition of soil organic matter (SOM) (Table 2.1). These environmental effects modify the temperature sensitivity of

Table 2.1   Factors affecting the temperature sensitivity of SOM decomposition (Davidson and Janssens 2006).

Physical Protection	OM within soil aggregates is physical protected from microbial enzyme action
Chemical Protection	OM adsorbed onto soil mineral material is chemically protected from decomposition
Drought	Inhibits the movement by diffusion of enzymes and substrates
Flooding	Soil saturation leads to anaerobic conditions
Freezing	Freezing limits diffusion of substrates and enzymes

OM decomposition rates, which are also controlled by the nature of the substrate (complex substrates have a higher sensitivity because they have higher activation energies) (Figure 2.7). The environmental controls impact on the transport of enzymes and substrate within the soil, so that rates of decomposition by microbially derived extracellular enzymes are reduced. The mean residence time (MRT) of soil organic carbon (SOC) is defined as:

$$MRT = SOC / R_h$$

where R_h is the rate of heterotrophic respiration (Chen et al. 2013). MRT can be estimated from carbon stock measurement and measured rates of respiration but can also be approximated through radiocarbon aging of soil carbon (Shi et al. 2020) which is a valuable tool for assessing the impact of changing soil conditions on carbon storage or CO_2 release.

Geomorphological Controls on Heterotrophic Soil Respiration

In the context of geomorphological controls on soil organic matter decomposition, the role of physical and chemical protection is of particular interest, since it is dependent on interactions between soil organic matter and mineral matter. Consequently, erosion, transport and deposition of soils and sediments that produce mixing of organic and mineral matter are potentially important. Organic matter is stabilised either by becoming physically protected within mineral aggregates or through binding to mineral surfaces (particularly iron manganese and aluminium oxides and phyllosilicate minerals).

Davidson and Janssens (2006) distinguish between long-lived organic matter in soils which is 'stabilised' by binding to mineral material and organic matter which is 'preserved' in anaerobic conditions. Geomorphological processes of erosion, transport and burial of organic and mineral sediments might promote both of these mechanisms, but they may also excavate preserved organic matter that has been buried in the aerobic zone and may also lead to the physical disruption of soil

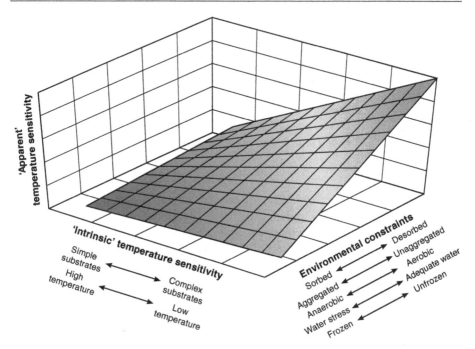

Figure 2.7   Controls on temperature sensitivity of SOM decomposition. Source: After Davidson and Janssens 2006. Reproduced with permission of Springer Nature.

aggregates. The net effect of erosion on soil organic matter preservation is therefore a subject of intensive research and debate (see Chapter 6), but the nature of environmental controls on soil respiration rates means that the impacts of redistribution of earth surface materials cannot be ignored in understanding spatial patterns of carbon sequestration in soil systems (cf. Berhe et al. 2018).

Organic Matter Quality, Soil Carbon Storage and Decomposition

Soil carbon storage is not only affected by simple physical controls on respiration and by microbial responses to these controls, but also by the nature and availability of soil organic matter, which has an important influence on the rates of carbon turnover in the soil system. Carbon storage in soils is the largest stock in the terrestrial carbon cycle, so that understanding of the nature of this material, and controls on its rate of turnover, are an important part of modelling the earth system.

   Organic carbon fixed in plant tissues by photosynthesis is transferred to the soil either as litterfall on plant senescence or through the release of root exudate directly to the soil. In the soil, microbial decomposition of organic matter produces three principal classes of product. One possible product is gaseous release of carbon occurring either as carbon dioxide (mineralisation due to aerobic respiration) or as methane (driven by methanogenic bacteria operating under anaerobic conditions). A second component is soluble humic and fulvic acids produced by enzyme catalysed hydrolysis of plant tissues. These are typically flushed to the drainage network by soil water movement, and along with soluble root exudates (carbohydrates), are the principle natural components of dissolved organic carbon (DOC). The third product of decomposition of soil organic matter is the solid residue of organic matter. This component is formed of larger particles and of more biochemically stable compounds that are relatively resistant to microbial metabolisation. These elements of the SOM are described as recalcitrant.

The conceptualisation of soil organic matter as transitioning from more labile constituents to more recalcitrant material is a standard approach in soil science (e.g., Figure 2.8). This concept underpins most approaches to modelling the behaviour of organic matter stored in soil systems. Figure 2.8 represents the range of organic matter states in soils as a continuum, but they are commonly modelled as a series of soil carbon 'pools'; effectively modal states of organic matter with characteristic turnover times (e.g., Table 2.2).

The classification represented in Table 2.2 is a pragmatic classification of organic matter types and turnover times, but conceptually it is not particularly coherent. The pools include two types of 'fresh' plant material (structural and metabolic), partially decomposed plant material (slow), microbial products (active) and the 'passive' pool, which may either be very recalcitrant organic matter, or more typically organic matter bound to mineral materials. Arguably, the main theoretical distinction here is between mineral-bound and non-mineral bound organic matter, with the latter spanning a gradient from recalcitrant to labile as a function of molecular and structural composition. This complexity means that it can be difficult to link the conceptual representation of carbon pools with empirical measurement of soil carbon in the laboratory. A range of approaches have been applied; for example, Skjemstad et al. (2004) calibrated three fractions of the Roth-C model (resistant, inert and humic organic matter) by measuring particulate organic carbon, charcoal and humic substances within soil samples.

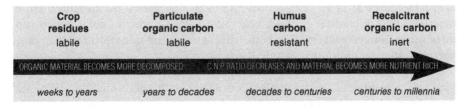

Figure 2.8 SOM decomposition conceptualised as metabolisation of material of increasing recalcitrance at increasing timescales. Source: After soilquality.org.au. Courtesy of Ram Dalal.

Table 2.2   Soil carbon pools as characterised in the Century soil carbon model.

Soil Carbon Pool	Source	Turnover time
Active	Labile microbial products	1–5 years
Slow	Less labile SOM	25 years
Passive	Recalcitrant material or bound to mineral surfaces	200–1500 years
Structural	Plant residue	1–5 yr
Metabolic	Plant residue	0.1–1 yr

Source: After Parton et al. 1987.

Some recent work has questioned the conceptualisation of organic matter recalcitrance as a control on the eventual storage or mineralisation of soil organic matter. Whilst it is the case that simple organic molecules such as carbohydrates are more directly available to support microbial metabolism fungal and bacterial extracellular enzymes are capable of breaking down more complex organic molecules (such as lignin and cellulose) to easily metabolised constituents. Lützow et al. (2006), in a comprehensive review of organic matter decomposition, suggest that recalcitrance is only relative, and that soil biota can decompose any organic matter given enough time. Burns et al. (2013) describe the ways in which 'cocktails' of extracellular enzymes are derived from diverse populations of microorganisms, which exist across microsites, where physical conditions may vary at scales of nanometres to centimetres. This diversity of extracellular enzymes produced from fungi, bacteria and archaea living in soil waters and bound to soil constituents, produces an aggressive environment which rapidly decomposes the full spectrum of organic molecules in soil systems. In this understanding, SOM decomposition is a 'community driven process' (Wallenstein and Burns 2011), so that environmental responses of SOM rates are complex and influenced by the autecology of the range of decomposer microorganisms. In favourable oxic conditions, whilst 'recalcitrant' organic matter may take longer to decompose (years to decades), long-term preservation of OM is more likely to be due to chemical stabilisation than preservation of recalcitrant material. This conceptualisation of SOM decomposition is becoming more widely accepted (e.g., Marschner et al. 2008; Schmidt et al. 2011). Hemingway et al. (2019) present global data based on measurement of organic matter activation energies and radiocarbon dating of soils, which supports much longer preservation times (millennia) for mineral bound organic matter relative to unbound organic matter in soils.

Commonplace observations such as the paucity of organic matter preservation in semi-arid soils are also consistent with this understanding, and phenomena such as 'priming' of OM decomposition further support the idea that recalcitrance is a relative term. Priming is the term used to describe the observations that addition of labile compounds to systems can promote the decomposition of apparently recalcitrant material (van der Wal and de Boer 2017). This phenomenon is commonly understood to be a result of the use of labile material by microorganisms as a substrate, supporting the generation of extracellular enzymes, which then degrade larger and more recalcitrant molecules. van der Wal and de Boer (2017) argue that the strength of priming is related to the existence of synergistic combinations of the priming compound, the nature of existing SOM, and the potential of the microbial communities to degrade that OM. This means that where the physico-chemical soil conditions are not limiting, and where suitable microbial communities can access the organic substrate, decomposition of recalcitrant material is to be expected.

These observations have significant implications for our understanding of carbon sequestration in soil and sediment systems. Decomposition of organic matter in soil and sediment systems is a function of a combination of interacting controls which include the microbial community, the nature of the substrate, physical and chemical conditions at microsites in the soil system and interactions between organic and mineral sediments (Figure 2.9). However, the probable fate of organic matter added to soil systems is decomposition and respiration to CO_2. Long-term sequestration of organic matter may occur primarily under two

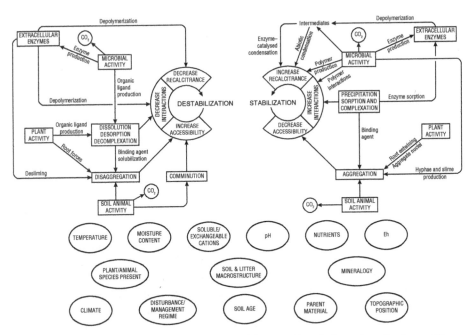

Figure 2.9   Summary of controls on the stabilization and decomposition of soil organic matter. Source: After Sollins et al. 1996. Reproduced with permission of Elsevier.

conditions: 1) where OM is stabilised through interactions with mineral grains; and 2) where persistent saturation produces consistently anaerobic conditions, e.g., in wetland soils. These are physically determined characteristics of soil and sediment systems, which have the potential to be controlled and mediated by geomorphological processes.

Geomorphology and Carbon Exchange from Soils and Sediments

On the basis of the understanding of the controls on primary productivity and organic matter decomposition discussed above, geomorphological controls on terrestrial carbon storage can be classified into four main elements:

1) Topographic control on carbon cycling processes – topography impacts on soil respiration through impacts on temperature and soil moisture and indirectly through the impact of relief on the physical stability of soil and sediment carbon stores. Similarly, changes in temperature and moisture impact on vegetation communities and thus on primary productivity (e.g., Mildowski et al. 2015).
2) The impact of geomorphological processes on primary production – in areas of high geomorphological activity, high rates of erosion, transport and deposition of sediments effect carbon cycling. Geomorphological disturbance of vegetation communities (e.g., by flooding or landslide disturbance) leads to primary succession which can enhance primary productivity (Swanson et al. 2010). Rates of erosion can also influence weathering rates and consequent nutrient availability (Porder et al. 2007, 2015). Nutrient status is further modified through lateral flux of material leading to local enhancement or depletion of soil nutrient status (Berhe et al. 2018) increasing or reducing productivity.
3) The impact of geomorphological processes on decomposition of SOM – mobilisation of sediments by geomorphological processes can also amplify mixing of organic matter and fresh mineral surfaces, escalating stabilisation of organic matter, but it also has the potential to modify local biogeochemical environments, for example by eroding previously stored carbon and re-exposing it to aerobic processes at the earth's surface. Oxidation and burial of eroded carbon at hillslope scale is a well-studied area with potential to significantly impact terrestrial carbon budgets (Berhe et al. 2018).
4) Landforms as sites of carbon storage – the suite of geomorphological processes operating within a catchment (partially conditioned by topography), the rates at which they operate, and their spatial configuration define the landscape sediment budget. This in turn defines the spatial pattern of sediment accumulation and erosion, which produces characteristic suites of erosional and depositional landforms within a landscape. Depositional landforms are potential sites of carbon sequestration and turnover. The volume of these material stores is a

physical limit on carbon storage, and the rate of turnover of sites (formation and erosion of landforms at annual to millennial timescales) is an important control on rates of carbon turnover in the sediment system.

In addition to the 'at a point in space' impacts of geomorphological change in controlling vertical carbon exchange between the terrestrial system and the atmosphere, the material fluxes driven by geomorphological processes also directly transport organic carbon through the terrestrial system. These fluxes are large and globally significant but have been poorly recognised in models and analysis of terrestrial carbon cycling. They are considered in more detail below.

Lateral Transfers of Carbon in the Fluvial System

Studies of terrestrial carbon cycling have been dominated by ecological research and have typically focused on carbon sequestration understood as the balance of NPP and soil respiration, as outlined above. Consequently, the focus has been on 'vertical' fluxes of carbon from the biosphere to the pedosphere, and on fluxes of methane and carbon dioxide to the atmosphere. Studies of lateral carbon fluxes have focused on estimates of land-ocean fluxes of dissolved and particulate carbon measured at river outlets (Schlesinger and Melack 1981), and have been derived either from extrapolations of increasingly large global discharge and carbon concentration datasets, or have been modelled from catchment-scale predictors of fluvial carbon flux (Table 2.3). Estimates from a range of studies over the past three decades have been relatively stable, so that a reasonable estimate is that global fluvial carbon flux to the oceans is

Table 2.3 Estimates of global fluvial carbon flux to the ocean in PgC per year.

DOC	POC	DIC	PIC	Notes	Reference
0.25	0.15				Hedges et al. 1997
				Extrapolated from 12 global rivers	Schlesinger and Melack 1981
0.22	0.18	0.17	0.38		Meybeck 1982
0.2	0.1	0.24			Meybeck 1993
0.36				Predicted based on relation with soil C:N ratio	Aitkenhead and Mc-Dowell 2000
0.17				Based on 118 world rivers	Dai et al. 2012
0.24	0.24	0.41	0.17		Li et al. 2017a
0.21	0.19	0.32		Modelled from 29 rivers	Ludwig et al. 1996a

in the range 0.8–1.3 PgC per year (Huang et al. 2012). Four components of fluvial carbon flux are commonly measured, DOC, dissolved inorganic carbon (DIC), particulate organic carbon (POC) and particulate inorganic carbon (PIC), although not all studies have considered all components. The most commonly measured components in relation to the fast carbon cycle are DOC and POC, which are present in approximately equal proportions in the global fluvial load (~ 0.2 PgC/yr each; Table 2.3).

Within these global loads, significant spatial variability is observed. Dai et al. (2012) have shown that the largest fluxes of DOC are to the Atlantic and western ocean boundaries, with the impact of large Amazonian fluxes producing a high overall flux from South America (Figure 2.10). Figure 2.10 also demonstrates the dominance of river discharge as a control on DOC flux. Global mapping of fluvial carbon yield by Li et al. (2017a) (see Figure 2.11) shows similar patterns, but also emphasises the significance of high northern latitudes, where organic rich soils of the northern boreal and tundra zones produce high areal DOC and POC fluxes (cf. McClelland et al. 2016). However, in terms of absolute global carbon flux to the ocean, lower latitudes dominate, with Dai et al. (2012) estimating that 62% of total fluvial DOC flux is derived from latitudes 30°S to 30°N.

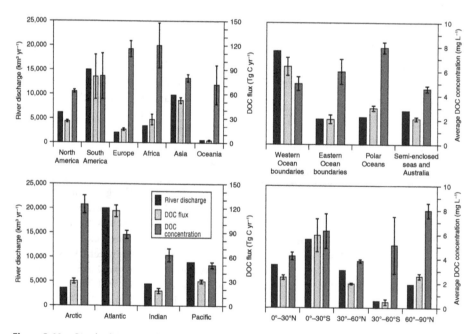

Figure 2.10   Dissolved organic carbon flux to the oceans. Source: After Dai et al. 2012. Reproduced with permission of Elsevier.

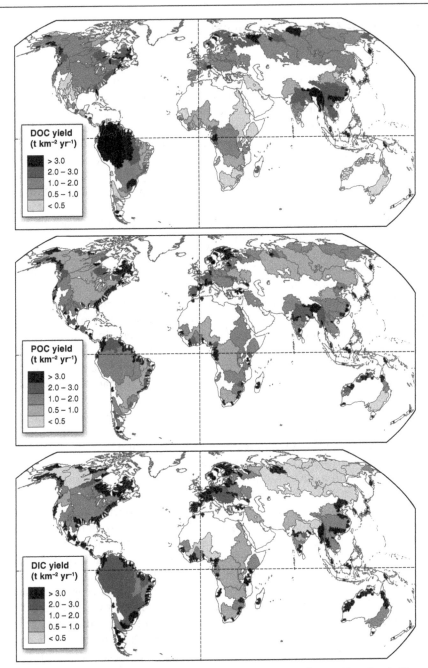

Figure 2.11   Global patterns of carbon flux to the oceans. Source: After Li et al. 2017a. Reproduced with permission of Elsevier.

Controls on the Geography of Fluvial Carbon Flux

Analysis of the spatial patterns of fluvial carbon flux has the potential to develop understanding of large-scale controls on terrestrial carbon flux to the ocean. Controls on spatial patterns of terrestrial carbon delivery include factors deriving from the source of that carbon (i.e. the nature of carbon fixed by the biosphere and the decomposition of that material), as well as the processes which drive the physical transport of material from source to sink. Aitkenhead and McDowell (2000) showed that based on 164 rivers, division of the land surface into 15 global biomes allowed fluvial DOC flux to be predicted from the C:N ratio of the catchment soils (accounting for 99% of observed variance in flux). Despite the very strong predictive power of this relation, the mechanism linking C:N ratios to terrestrial fluvial carbon flux is not well understood. Aitkenhead and McDowell speculate that they are linked through a common dependence on temperature and precipitation (which are key controls on biologically mediated production of DOC), on catchment runoff, and on biological production in catchments.

Li et al. (2017a) note the association of peat covered catchments, together with high sediment yields and high runoff per unit area in tropical regions, as a key driver of high fluvial carbon fluxes in these areas. The key control of discharge on carbon fluxes has been noted in a range of contexts, including blanket peatlands (Gibson et al. 2009) and montane forests (McDowell and Asbury 1994). This widespread association suggests that in many systems, DOC export is effectively transport limited, so that the sensitivity of the flux to changes in discharge is greater than to changes in the rate of DOC production in soils.

Clair et al. (1994) showed that in rivers draining the eastern Canada basin, area and topography were key predictors of DOC flux, and suggested that lower basin topography was associated with greater wetland areas and therefore higher DOC flux. Similarly, Ludwig et al. (1996a) assessing global inputs of organic carbon to the ocean based on a dataset of global rivers, showed that catchment slope, discharge and the carbon content of catchment soils were the strongest predictors of organic carbon flux. Current evidence therefore supports a strong association between highly organic soils and high rates of runoff production, resulting in an elevated DOC flux to the oceans. These are characteristics of wetland soils, which is apparent from the coincidence of high DOC fluxes in Figure 2.11a, and areas of extensive peatlands in the high latitudes and in tropical regions.

Global patterns of particulate carbon loss from terrestrial ecosystems suggest two potential controls on the magnitude of the POC flux to the oceans (Figure 2.11b). High POC fluxes are associated with regions of high carbon storage in the biosphere, in particular soil systems such as the high latitude peatlands and the Amazon basin. However, high rates of POC flux are also associated with areas of elevated sediment flux, such as the western cordillera of North America and the Himalayas. Galy et al. (2015) estimated fluxes of biospherically derived POC and petrogenic POC based on data from 70 large river systems. Petrogenic POC is fossil organic carbon that has been stored in rocks and sediments, and any fluxes are strongly driven by physical erosion, so that fluxes of POC correlate strongly with suspended sediment yields. Galy et al. show that fluxes of biosphere-derived POC (largely fresh plant matter and plant matter preserved in and then eroded from soil organic horizons) are also strongly controlled by erosion rates. Biospheric POC export correlates only weakly with NPP, indicating that export of particulate carbon is a transport limited process. Globally, Galy et al. demonstrate that even at the highest erosion rates, the proportion of NPP which is exported annually is less than 3% and for most systems it is less than 1%. Therefore, whilst DOC flux appears to be controlled both by soil carbon concentrations and by drivers of fluvial flux, POC is primarily controlled by transport limited processes. Changes in rates of soil erosion and landsliding, and tectonic and climatic controls on rates of continental denudation, are therefore important controls on POC flux to the oceans. Galy et al. argue that because rates of carbon burial tend to rise with an increase in sediment yield to the ocean (e.g., Hilton et al. 2011a), accelerated rates of continental erosion lead to drawdown of CO_2 from the atmosphere.

In the context of the wider carbon budget, ocean burial of carbon is a relatively small flux. POC and DOC delivered to the ocean are rapidly oxidized with an estimated 55–80% of terrestrial organic carbon flux oxidized along the continental margin (Blair and Aller 2012). Modern biomass is labile and rapidly decomposed, whereas fossil material is more recalcitrant. Along active plate margins, where erosion rates are higher, fossil material is a higher proportion of total load and there is more opportunity for rapid burial of modern carbon so that ocean burial of terrestrial carbon is concentrated in these zones (Blair and Aller 2012). It is important to note that burial of modern biomass represents a removal of carbon from the biosphere–atmosphere system at timescales comparable to changes in atmospheric carbon concentrations, whereas burial of fossil carbon represents a transfer from one long-term store to another.

Carbon Transformations in Freshwater Systems

Fluvial carbon is rapidly metabolised in estuaries and oceans, but it had been widely assumed that the relatively short residence times of carbon within the fluvial system meant that carbon transport within rivers was a relatively conservative system. In the last decade, the idea that river systems are passive conduits transporting terrestrial carbon to the ocean has been significantly challenged (Aufdenkampe et al. 2011; Battin et al. 2009; Cole et al. 2007). Cole et al. (2007) conservatively estimate that 1.9 PgC per year were added to inland waters from terrestrial ecosystems, of which just 0.9 PgC is delivered to the oceans. Battin et al. (2009) argued that carbon is sequestered, transported and transformed in freshwater systems, so that the flux of carbon to the ocean measured at river outlets is only a small proportion of the carbon exported from the terrestrial system to freshwaters. Further work by Aufdenkampe et al. (2011) has estimated that of 2.7 PgC exported to freshwater systems, the flux to the ocean is only 0.9 PgC, with 1.2 PgC lost to the atmosphere through mineralisation and 0.6 PgC sequestered in lake, reservoir and wetland systems.

The understanding that carbon transformations can occur within the fluvial system has partly been driven through the recognition of processes which lead to rapid cycling of fluvial carbon at timescales relevant to the residence time of river waters. For example, Moody et al. (2013) estimate in-stream losses of up to 69% of dissolved carbon in the River Tees, driven by rapid mineralisation of DOC by UV light. However, equally important has been recognition that the residence time of the fluvial carbon load is not necessarily constrained by the mean travel time of river waters from source to sink, but that temporary storage in depositional zones means that significant carbon cycling occurs within the fluvial system.

Battin et al. (2008) argued that downstream changes in channel hydraulics lead to variation in microbial niches (Figure 2.12). They argue that geophysically determined sites of microbial carbon metabolism shift downstream, from biofilms in headwaters, to suspended aggregates and floodplain sites in the lower course. In particular, floodplain storage and metabolism of dissolved and particulate carbon is important because of the much longer residence times of carbon in these depositional environments (at least the period between flood pulses). The role of floodplains in terrestrial carbon cycling is considered in detail in Chapter 7.

Insights into the importance of carbon storage and remobilisation within the fluvial system can be gathered by considering the age of the carbon transported. Radiocarbon aging of dissolved and particulate organic matter demonstrates that

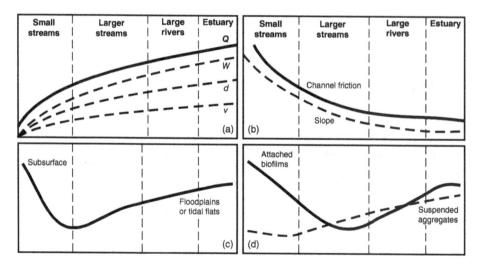

Figure 2.12   Changing microbial niches through river system networks controlled by changing channel geomorphology and hydrology. Source: After Battin et al. 2008. Reproduced with permission of Springer Nature.

whilst most of the fluvial carbon load is young carbon derived from rapid cycling of terrestrial organic matter delivered to headwater streams (Battin et al. 2009; Evans et al. 2014a; Moore et al. 2013), older radiocarbon dates indicative of the remobilisation of old carbon sequestered in catchment soils and sediment systems have been observed from rivers across the globe.

In the Ganges system, Galy and Eglinton (2011) report radiocarbon ages of biospherically derived POC of up to 16 800 BP, with the majority of samples in the range 460–5000. Similarly, on the Mackenzie River in northern Canada, Hilton et al. (2015) report mean radiocarbon ages for biospherically derived POC of 5800 BP.

Old particulate carbon is released to the fluvial system from storage in catchment soils and sediments due to erosion. This may be natural or linked to human disturbance. Human disturbance can also return organic matter to the oxic zone from burial in anaerobic sediments due to drainage, erosion or disturbance where it may be decomposed releasing dissolved carbon. Butman et al. (2014) analysed a global dataset of fluvial radiocarbon ages and demonstrated a linear association between indices of catchment disturbance and increasing age (Figure 2.13). They estimate that globally 3–9% of DOC in river systems is derived from aged carbon sources due to human disturbance.

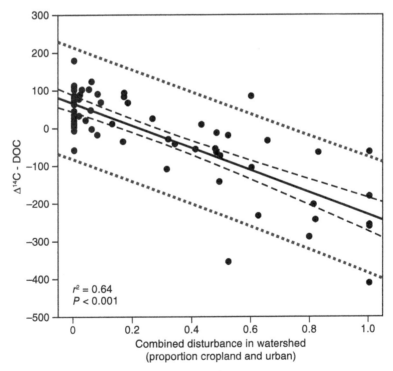

Figure 2.13   Global dataset illustrating older radiocarbon 'ages' associated with DOC derived from highly disturbed catchments. Source: After Butman et al. 2015. Reproduced with permission of Springer Nature.

Carbon Turnover in the Fluvial System

What is becoming increasingly apparent is that fluvial systems are dynamic sites of carbon storage and/or carbon turnover. Carbon is stored in depositional land-forms within catchment systems and released through hillslope or fluvial erosion of these sites. Organic matter is delivered to river systems through erosion (POC) or dissolved in hillslope runoff, where it is actively cycled under control of both physical and biological processes. The fluvial geomorphology of a river system determines the physical template within which microbial turnover of carbon occurs. The recognition that transformation of carbon between the solid, dis-solved and gaseous phases occurs within the river system is fundamental to one of the most significant research challenges in our understanding of terrestrial carbon cycling. This is the question of the fate of organic matter mobilised within catchment systems. The carbon in this organic matter may enter long-term storage in soils and sediments within the catchment system, it may be miner-alised from sites of shallow soil storage, it may be transformed and mineralised in transport in the fluvial system, or it may be delivered to the ocean where it is most likely to be rapidly oxidised, but a proportion of this material will be bur-ied and enter long-term carbon storage. The relative importance of preservation, transformation and mineralisation of carbon in catchment systems is discussed in more detail in the following chapters.

Chapter Three
Geomorphology and the Geological Carbon Cycle

The Geological Carbon Cycle

In the preceding chapter, the fixation and turnover of organic carbon in the terrestrial carbon cycle was discussed, and the links between geomorphological processes and carbon turnover were initially explored. Carbon is also cycled through the earth system at much longer timescales. The geological carbon cycle, sometimes described as the slow carbon cycle, links terrestrial weathering and oceanic sedimentation (Berner 2003). These processes are intimately associated to key areas of geomorphological study, with patterns of erosion and deposition being fundamental to analysis of the system. The key interactions of the geological cycle with the carbon cycle can be characterised according to a series of linked chemical reactions (Hilton and West 2020; Sundquist 1991), as outlined in Table 3.1.

Weathering of silicate minerals occurs in the presence of atmospheric CO_2 in the terrestrial environment and produces dissolved ions that are transported to the oceans. Here, sedimentation of carbonates, largely through biological fixation and deposition on the sea bed, transfers carbon which has been removed from the atmosphere to long-term storage in the lithosphere (Ruddiman 2008). Carbonate deposition is constrained to ocean waters where the bed is above the carbonate compensation depth (CCD) (typically 4–5000 m), below which high pressure, low temperatures and elevated CO_2 concentrations increase calcite dissolution to rates above that of the supply of carbonate from the shallower ocean.

Geomorphology and the Carbon Cycle, First Edition. Martin Evans.
© 2022 Royal Geographical Society (with the Institute of British Geographers). Published 2022 by John Wiley & Sons Ltd.

Table 3.1   Summary chemistry of the main processes in the geological carbon cycle.

Silicate Weathering

$3H_2O + 2CO_2 + CaSiO_3$	$\rightarrow$	$Ca^{2+} + 2HCO_3^- + Si(OH)_4$
$3H_2O + 2CO_2 + MgSiO_3$	$\rightarrow$	$Mg^{2+} + 2HCO_3^- + Si(OH)_4$
$H_2O + CO_2 + CaCO_3$	$\rightarrow$	$Ca^{2+} + 2HCO_3^-$

Sedimentation

$Ca^{2+} + 2HCO_3^- + Si(OH)_4$	$\rightarrow$	$CaCO_3 + SiO_2 + 3H_2O + CO_2$
$Mg^{2+} + 2HCO_3^- + Si(OH)_4$	$\rightarrow$	$MgCO_3 + SiO_2 + 3H_2O + CO_2$
$Ca^{2+} + 2HCO_3^-$	$\rightarrow$	$CaCO_3 + H_2O + CO_2$

Decarbonation

$CaCO_3 + SiO_2 + 3H_2O + CO_2$	$\rightarrow$	$3H_2O + 2CO_2 + CaSiO_3$
$MgCO_3 + SiO_2 + 3H_2O + CO_2$	$\rightarrow$	$3H_2O + 2CO_2 + MgSiO_3$

Sulphide Oxidation

$4FeS_2 + 15O_2 + 14H_2O$	$\rightarrow$	$4Fe(OH)_3 + H_2SO_4$
$FeS_2 + 14\,Fe^{3+}\,8\,H_2O$	$\rightarrow$	$15Fe^{2+} + 2SO_4^- + 16H^+$
$CaCO_3 + H_2SO_4$	$\rightarrow$	$CO_2 + H_2O + Ca^{2+} + SO_4^{2-}$

Oxidation of Fossil Organic Carbon

$CH_2O + O_2$	$\rightarrow$	$CO_2 + H_2O$

Source: After Hilton and West 2020; Sundquist 1991.

Below the CCD, carbonate is not deposited in ocean sediments (Ridgwell and Zeebe 2005).

Two CO_2 molecules are consumed in the silicate weathering reaction, but only one is released during sedimentation, so that the geomorphological processes of erosion and deposition of silicate minerals potentially control rates of drawdown of CO_2 from the atmosphere. In contrast, carbon dioxide consumption and production during the cycle of weathering and sedimentation of carbonate rocks does not impact on atmospheric carbon concentrations, since the single molecule of CO_2 consumed in the weathering reaction is released during sedimentation. This has important implications for assessment of weathering rates in the context of carbon cycling, as total ionic concentrations in runoff do not simply represent weathering reactions that draw down CO_2 from the atmosphere.

At much longer timescales, heating of carbonate rocks through deep burial or tectonic activity leads to decarbonation releasing CO_2, which is returned to the atmosphere largely through degassing from volcanic vents (Figure 3.1). The carbon balance across the cycle is therefore maintained over longer timescales because one molecule of CO_2 is evolved during decarbonation (Penman et al. 2020). However, the processes and timescales of erosion and sedimentation are

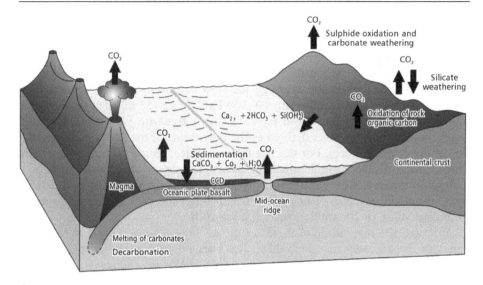

Figure 3.1   The geological (slow) carbon cycle. Source: Martin Evans.

distinct from the controls on decarbonation identified above, so that net draw-down of CO_2 from the atmosphere due to erosional processes is possible.

The interactions of the silicate weathering cycle with oceanic carbon cycling have been the main focus of the scientific understanding of the slow carbon cycle but more recently the role of weathering of organic rich sedimentary rocks and sulphide minerals in carbon cycling has also been explored (Hilton and West 2020). Oxidation of fossil organic carbon preserved in sedimentary rocks releases carbon dioxide directly from the lithogenic store to the atmosphere (Petsch 2014) as a result of microbial respiration of fossil carbon sources (Hemingway et al. 2018; Keller and Bacon 1998). Weathering of sulphide minerals which are also concentrated in organic rich sedimentary strata releases sulphuric acid. When carbonate minerals react with sulphuric acid, one molecule of carbon dioxide is released so that sulphide mineral weathering also leads to carbon flux to the atmosphere (Torres et al. 2014).

The aim of this chapter is to assess firstly, the impact of geomorphological processes on the operation of the geological carbon cycle, and secondly, ways in which the conceptual understanding of terrestrial sediment systems which geomorphologists have developed can inform this analysis. The focus here is pri-marily on terrestrial processes. Detailed consideration of oceanic carbon cycling is beyond the scope of this chapter, because the impact of geomorphological processes on the oceanic system is largely secondary, relating to changing terres-trial inputs of organic and inorganic carbon to the oceanic system.

Rates of Geological Carbon Cycling

The magnitude of flux in the slow carbon cycle is relatively small in comparison to rates of cycling of carbon through the fast carbon cycle (Figure 3.2). On an annual basis, the carbon flux through the slow carbon cycle is on the order of 0.01–0.1 PgC a^{-1}, contrasting with 10–100 PgC a^{-1} cycled through the fast carbon cycle (NASA 2016). These reduced fluxes translate into lower turnover times (with carbon turnover defined as the magnitude of the flux, divided by the size of the relevant store). For the fast carbon cycle, turnover times are typically on the order of years to millennia. In contrast, the slow carbon cycle turns over carbon on timescales of tens of millennia to millions of years (Ciais et al. 2013), with complete recycling of oceanic crust through the rock cycle occurring at timescales of up to 200 million years (Grace 2001).

Although the magnitude of fluxes in the geological carbon cycle is relatively small, the timescales over which these fluxes are relevant are long, so that if the cycle were not in balance, it would drive very significant changes in long-term

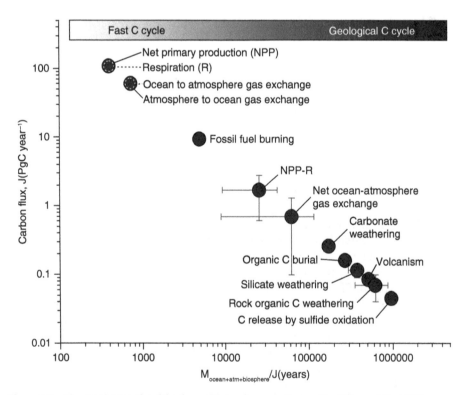

Figure 3.2   Fluxes and timescales of the slow and fast carbon cycle. Source: After Hilton and West 2020. Reproduced with permission of Springer Nature.

atmospheric CO_2 concentration. For significant changes not to occur, there must be a balance of CO_2 sequestered by weathering and ocean sedimentation and the release of CO_2 from decarbonation of lithified carbon and oxidation of rock derived organic carbon. Considerable uncertainties are associated with balancing the contemporary geological carbon cycle. However, as noted by Kump et al. (2000), the magnitude of the fluxes are small compared to the ocean–atmosphere carbon store, so that the response time of the system is likely to be circa 105 years. Consequently, changes in the geological carbon balance are only likely to impact atmospheric CO_2 concentrations where they persist over geological time-scales (Kump et al. 2000).

The other side of this coin is the argument made by Berner and Caldeira (1997), who conclude that because the atmosphere-ocean-biosphere-soil store of global carbon is over three orders of magnitude smaller than the geological store, the balancing of global carbon cycling over the long term must be controlled by the slow carbon cycle exchanging carbon with the geological store through the processes of silicate weathering and decarbonation. Over long time periods, significant imbalance between these processes would lead to atmospheric carbon concentrations outside the accepted range for the past 600 million years (Figure 3.3).

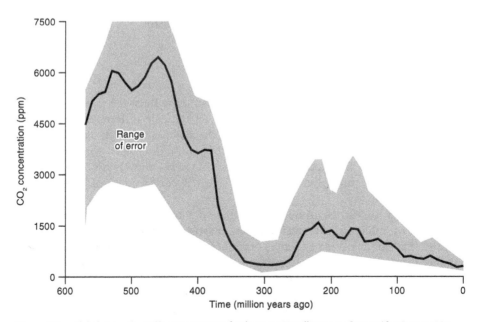

Figure 3.3   Global atmospheric CO_2 concentrations for the past 550 million years. Source: After Berner 1997. Reproduced with permission of The American Association for the Advancement of Science.

The idea of the geological carbon cycle as a self-limiting system which acts as a 'global thermostat' (Archer 2008) was first advanced by Walker et al. (1981), who argued that the rate of silicate weathering is controlled by temperature, and that because temperature is controlled by atmospheric CO_2 concentrations, high values will be buffered by accelerated rates of weathering (and so greater removal of CO_2 from the atmosphere) and lower values by reduced weathering. Zeebe and Caldeira (2008) provided support for the role of weathering in global homeostasis by using ice core records to constrain models of carbon cycling over the last 610 000 years. They demonstrate that over this period, the maximum difference between uptake of CO_2 by weathering and release by decarbonation is 1–2%, arguing that this is evidence for a highly effective weathering feedback loop. Berner and Caldeira (1997) suggest from the work of Sundquist (1991) that the lag time required for this mechanism to return the global system to a steady state is in the order of 106 years. This figure is also indicated by modelling results, as the amount of time for the weathering system to mitigate contemporary increases in atmospheric CO_2 (Lenton and Britton 2006).

Recognition that weathering of organic rich sedimentary rock can release CO_2 to the atmosphere complicates the traditional view that weathering draws down atmospheric carbon and these ideas have not been fully integrated into the paradigm of the self limiting geological carbon cycle. Estimated rates of CO_2 release from weathering of rock derived organic carbon are comparable to rates from volcanic degassing. As Hilton and West (2020) argue, there is a need to develop carbon cycle models which incorporate these mechanisms.

Controls on Weathering Rates

Controls on the weathering feedback mechanism have been the subject of some controversy. One school of thought focusses on the role of temperature as the driving force behind a global thermostat as discussed above. An alternate view has developed from the work of Raymo and Ruddiman (1992), who argue that the principle driver of changing weathering rates over time is not temperature changes related to atmospheric CO_2, but relates to continental uplift in tectonically active areas.

Raymo and Ruddiman argued that global cooling in the last 40 million years was driven by enhanced chemical weathering rates due to Himalayan uplift. They propose that uplift has led to larger amounts of precipitation and consequently enhanced weathering rates. They further argue that glaciation is a positive feedback on this global cooling mechanism, since the production of fresh mineral material by glacial erosion will further accelerate weathering rates. One of the challenges of this hypothesis is the need for a negative feedback to prevent runaway cooling. Raymo and Ruddiman acknowledge this and appeal to changes in the rate of burial of organic carbon in the ocean, which have reduced by 50% over the Cenozoic era, although the nature of the link to temperature is not well defined.

In a recent compilation of global data on weathering, erosion and sedimentation, Willenbring and Jerolmack (2016) argue that data for the late Cenozoic indicates stability in global rates of landscape change with no observed trend in rates of weathering or sedimentation. They infer that the control of weathering on temperature (due either to changes in climate or uplift) is undermined by these results. In contrast, Caves et al. (2016) argue that declining Cenozoic concentrations of atmospheric CO_2 are driven by changes in the feedback strength between silicate weathering and climate, rather than by changes in the rate of weathering alone. They argue that the change in sensitivity through the Cenozoic era is driven by changes in lithological reactivity at the earth's surface, as a consequence of uplift and glaciation which expose more reactive lithologies. Thus, a short-term increase in weathering rate driven by uplift reduces atmospheric CO_2 concentrations, but this causes cooling and reduced runoff leading to a reduction in silicate weathering rates.

One approach to determining landscape-scale controls on weathering has been to assess weathering rates based on concentrations of weathering products in fluvial runoff. Gaillardet et al. (1999) analysed a suite of data from the largest 60 rivers in the world and suggested that runoff, temperature and physical erosion controlled overall erosion rates, and that only active denudation maintains high rates of weathering. They also argued that basalt weathering on oceanic islands was disproportionately important because of high basalt weathering rates. Li and Elderfield (2013) have similarly shown a key control of basalt weathering on the geological carbon cycle. By modelling continental silicate weathering and island basaltic weathering separately, they show that that reduced basalt weathering over the last 100 million years is consistent with observed global cooling.

Millot et al. (2002) also exploited weathering flux data from fluvial systems in an analysis of chemical weathering fluxes in runoff in the Canadian Shield. Comparing these data with estimates from sites across the globe, they demonstrated a power law relationship between rates of physical and chemical denudation. The complexity of the linkages between site conditions, and physical and chemical weathering, mean that interpreting this relation as causal is somewhat speculative. However, it is consistent with the hypothesis that accelerated physical erosion drives rates of chemical erosion, through the exposure of fresh mineral surfaces to natural waters.

A more complete analysis of global controls on silicate weathering rates is offered by West et al. (2005). They argue that the rate of reaction in the weathering of silicate minerals by carbonic acid (Table 3.1) is theoretically controlled by supply of mineral material, water and acid, the reactivity of the minerals, and by temperature. Through a compilation of data describing catchment-based estimates of physical and chemical weathering, they demonstrate complex controls on weathering rates. They argue that at low rates of physical erosion, supply of mineral material is the limiting factor controlling weathering rates, whereas at higher erosion rates, mineral surfaces are in excess, so that weathering reactions become kinetically limited. In this case, runoff and temperature are key controls. This study did not explicitly consider soil depths; however, West (2012) modelled the depth of chemical weathering, showing that at low denudation rates, weathering of soil dominates, but that at higher rates, weathering fluxes increase because of greater bedrock contribution to total weathering flux. This has also been demonstrated at the local scale; for example, Larsen et al. (2014) have demonstrated that at sites in the southern Alps of New Zealand, rapid soil formation occurs at sites with the highest erosion rates. West argues that because the effects of temperature and run-off as controls on weathering flux are less at low denudation rates, surface erosion is an important contributor to the role of weathering as a global 'thermostat'.

Although the basic kinetics of the weathering equations identify the ulti-mate controls on weathering rates, the ways in which these are expressed in the physical environment are complex. Maher and Chamberlain (2014) modelled weathering using an approach where hydrological controls on weathering are more important than temperature. In this model, hydrology was represented by catchment travel times, and maximum weathering rates were derived when flow was sufficient to remove weathering products, but much lower where travel times were short relative to kinetic reaction rates. Like the work of West et al. (2005), these results point to the importance of fresh substrate to drive weathering, and Maher and Chamberlain argue that their results explain correlations between weathering and erosion, but are also consistent with a negative feedback between erosion and temperature.

The association of erosion rates (which drive exposure of fresh substrate) and rates of weathering is significant when considering the impact of weathering on the oxidation of organic carbon from sedimentary rocks. Hilton and West (2020) suggest that human-induced change in erosion rates has the potential to signifi-cantly increase CO_2 flux from this source, so that erosion related release of car-bon from lithogenic stores has the potential to exacerbate short-term increases in atmospheric carbon concentrations associated with fossil fuel combustion.

There is general acceptance that the geological carbon cycle is a fundamental control on the evolution of long-term global climate; however, it is clear from the above that the details of these interactions are contested. The importance of temperature, runoff, uplift and erosion as controls on weathering are common ground, but their relative importance and modes of interaction are an active field of research. Recognition of the role of weathering and erosion in the oxidation of organic rich sedimentary rocks means that that transfers of carbon from the lith-osphere to the atmosphere need to be considered when assessing carbon cycling at human as well as at geological timescales.

Geomorphological Controls on the Slow Carbon Cycle

Considerations of the impact of weathering on the global carbon cycle are commonly based on global modelling approaches with limited consideration of geography. Two recent reviews point to important ways forward to address this gap. In a review of the role of mountain building and erosion in the transfer of carbon from the geological store to the atmosphere, Hilton and West (2020) demonstrate that the interaction of erosion and lithology can produce differential carbon fluxes to the atmosphere. Uplift and consequent erosion in systems with extensive sedimentary lithology have the potential to enhance weathering of sulphide and organic-rich facies leading to carbon release to the atmosphere rather than the drawdown that an understanding of silicate weathering might predict. Similarly, in their review of interactions of geomorphology with the geological carbon cycle, Goudie and Viles (2012) argue that resolving understanding of this element of the global system requires understanding of the diversity of landscapes, and of the interactions of weathering and erosion. In an excellent review, Goudie and Viles conceptualise geomorphology as central to any analysis of weathering rates (Figure 3.4), and identify four main controls on weathering rates and consider where geomorphological knowledge can contribute to understanding of rates and processes (Table 3.2). What they essentially argue is that in the light of the demonstrable complexity of the weathering system, the difficulty of identifying consistent patterns in global weathering datasets is unsurprising. The kinetic

Table 3.2   Summarising key points from Goudie and Viles 2012.

Topic	Key Points
Tectonic controls on weathering	• Lithological controls particularly 5–10-fold variation in chemical weathering rates between acidic silicates and basalt • High weathering rates associated with rapid erosion in areas of tectonic uplift • Rapid weathering of volcanic islands
Climatic controls on weathering	• Spatial variation in weathering rates linked to geographic variation in runoff and temperature, and the role of soils and weathering mantle as controls on weathering rates • Dust deposition enhancing weathering rates through supply of fresh material.
Biological controls on weathering	• The role of vegetation cover in increasing soil CO_2 concentrations and enhancing weathering • The role of lichens in accelerating weathering on bare surfaces and promoting soil development • Enhanced weathering rates through the action of microorganisms.
Geomorphological controls on weathering	• The relationship between depth of regolith and weathering rate. Thin soils promote weathering but thicker weathering horizons produce lower rates by inhibiting water circulation, reducing contact between soil waters and fresh mineral material. • The positive relation between erosion rates and weathering (with evidence that this relation is reduced at the highest weathering rates) • The role of floodplains as potential foci for weathering.

and thermodynamic controls on weathering reactions are well understood, but the diversity of weathering environments where these reactions occur, means that there is considerable variability in the way rates of weathering are expressed in space. Goudie and Viles argue that 'a key research need is for more comparable datasets of chemical weathering rates from a wide range of geomorphic situations around the world', p. 69. The implication of this is that the box defining geomorphological impacts on weathering in Figure 3.4 should be drawn more widely. On the left-hand side of the diagram, 'Temperature', 'Water', 'Acids', 'CO_2' and critically 'Residence time' are identified as controls on chemical weathering. Together these define the weathering environment, and implicit in the quote above is that the weathering environment is closely correlated with geomorphic context. Depositional landforms are defined by the nature of the material, the mode of deposition and the time period over which deposition occurs, and so a combination of geomorphic context and climatic context define the weathering environment.

In traditional geomorphological thinking, weathering is typically seen as either a consequence of, or a precursor to, erosion, but the weathering of eroded material in secondary deposition sites (depositional landforms) has been a lesser focus. Bouchez et al. (2012) argue that the sedimentary load of large rivers is dominated by material from tectonically active mountain belts. This material does not move in a single event from source to ocean, but may be stored for long time periods in depositional environments on floodplains, and may be reworked

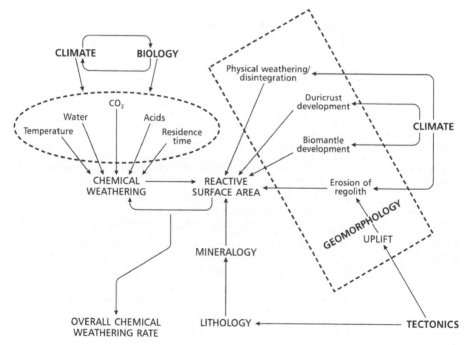

Figure 3.4   The role of geomorphology as a control on rock weathering rates. Source: After Goudie and Viles 2012. Reproduced with permission of Elsevier.

and redeposited several times before reaching the ocean. Consequently, there is considerable time for weathering of material in transit (Bouchez et al. suggest timescales of kyr to hundreds of kyr). Evidence from three river reaches in the Amazon basin, based on assessing weathering losses of dissolved Na, K, Mg and Ca, indicates that weathering of floodplain sediments may contribute up to 10% of the drawdown of CO_2 through chemical weathering in the river basins (Bouchez et al. 2012). Hilton and West (2020) similarly argue that depositional landforms such as floodplains and landslide deposits are 'weathering hotspots'. These locales can play an active role in carbon cycling, both through drawdown of CO_2 due to silicate weathering and release of CO_2 due to the weathering of rock derived organic carbon.

The role of floodplain sediments as weathering 'reactors' (Bouchez et al. 2012) is of particular interest in the context of the importance which has been ascribed to Himalayan uplift, weathering and erosion in global carbon cycling. Glacial erosion is a highly effective producer of fresh mineral surfaces through the mechanical grinding of rock to produce fine grained 'rock flour'. However, low temperatures minimise the degree to which this leads to accelerated weathering rates, so that measured rates of chemical weathering from glaciated and fluvial catchments in high mountains are similar (Anderson 2005). At large spatial scales however, the sediment cascade links the production of fresh mineral surfaces to conditions conducive to higher rates of chemical weathering (higher temperatures and abundant moisture) in lowland floodplains. In the Himalaya–Ganges system, West et al. (2002) studying a range of small catchments concluded that weathering rates on the Ganges plain are six times those observed in Himalayan catchments, with overall weathering rates comparable to those observed on basaltic islands. At a larger scale, Lupker et al. (2012) have shown depletion of mobile elements in floodplain contexts in the Ganges system, and similarly argue that the floodplain is the dominant location of chemical weathering. Frings et al. (2015) estimated weathering rates based on initial mobilisation of dissolved silica. They derived weathering rates for headwater Himalayan catchments twice that of lowland floodplains, but because of the greater area of the floodplain, the Ganges plain still showed that 41% of chemical weathering in the Himalaya–Ganges system is occurring on the floodplain compared with 25% in Himalayan systems. Although the weathering rates and relative weathering fluxes vary between studies, recent data are making it increasingly clear that the Gangetic floodplain plays a critical role in mediating the linkage between Himalayan erosion and accelerated weathering.

If it is recognised that the linkage between sediment sources and sinks which may be hotspots for weathering is an important part of developing an integrated view of global weathering, then it follows that it is also important to consider the temporal dislocation of erosion and associated weathering. Vance et al. (2009) developed a relation between weathering rates and age of weathered substrate, and demonstrated that in the present day, 18 kyr after the last glacial maximum, weathering rates are likely to be 2–3 times the long-term Quaternary average rate, due to the continued availability of relatively fresh mineral surfaces on fine-grained material derived from glaciation. Church and Slaymaker (1989) argued for an extended paraglacial effect apparent in sediment yield data from Canada, which implied that modern fluvial sediment yields are conditioned by the pulse of sediment released from melting glaciers at the end of the last glacial period, which is still being transported through the landscape. The observations of Vance et al. are a recognition that the large accumulations of paraglacial sediment which are in temporary storage in the world's major river systems impact on rates of chemical weathering, as well as the mobilisation of that material through the sediment cascade. Evidence of the impact of these processes on the Himalaya–Ganges system has been presented by Lupker et al. (2013), who have demonstrated that weathering proxies in sediments from the Bay of Bengal show evidence of elevated rates of chemical weathering since the Last Glacial Maximum.

Towards a Geomorphological Understanding of Global Weathering

The importance of considering secondary weathering environments dislocated from the locus of bedrock erosion, means that issues of scale are important to the understanding of the role of geomorphology in the geological carbon cycle. Analysis thus has tended to focus on large spatial scales, consistent with the large temporal scales that are relevant to the lithological cycling of carbon. Methodologically, approaches have included the analysis of continental scale fluxes of weathering products, the use of weathering proxies in ocean sediments, and global-scale modelling approaches linking carbon cycling and tectonic processes. However, Goudie and Viles (2012) also highlight the importance of small-scale process-based understanding of the weathering processes from the kinetics of chemical weathering under conditions of changing hydrology and temperature, to the small-scale action of lichens and microorganisms on rock surface weathering. Between the global scale and the scale of the microorganisms lies the bulk of the investigations of the earth's surface that geomorphologists have undertaken over the past century and a half (Figure 3.5).

Goudie and Viles conclude that the challenges of understanding the role of weathering in the geological carbon cycle relate to a lack of data collected at appropriate scales, and to the geographic complexity of the settings where weathering occurs. Weathering reactions are conceptually simple, with reaction kinetics controlled by the supply and removal of substrates and by temperature. The complexity arises because these reactions occur across a complex landscape, so that variation in space of temperature, substrate and the processes which mobilise

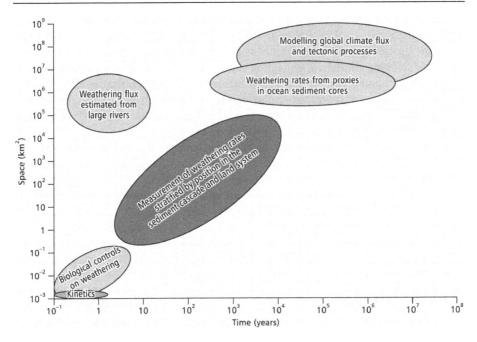

Figure 3.5 Time and space scales for weathering studies. The central space (darker grey shading) defined by timescales of circa 10–104 years and space scales of 1–10⁴ km has been less studied than larger and smaller scales, and at these landscape scales geomorphological expertise is particularly relevant in understanding controls on weathering systems. Source: Martin Evans.

weathering substrate and products leads to complexity of response. Developing an approach to analysis of temporal and spatial complexity in weathering regimes can usefully draw on the expertise of geomorphologists, who often work at intermediate scales, and on the suggestion from Goudie and Viles that 'Geomorphic situation' is a suitable sampling framework for weathering studies.

Church (2005) describes two divergent foci in contemporary geomorphology, a focus on earth system science, and process understanding of landscape systems at scales relevant to local environmental management. Researchers working on the first of these, most likely from earth science backgrounds, are usually working at larger time and space scales with a view to explaining global systems. Researchers in the latter area, more commonly from a geographical background, are more likely to work within a particular case study, or at the landscape scale, research that easily translates to the practical management of environments and landscapes. To date, most of the geomorphological insights into the debates around weathering, erosion uplift and the geological carbon cycle have been derived from the geological end of this spectrum. It is argued here, that to rise to the challenge posed by Goudie and Viles to incorporate geomorphological understanding of spatially variable weathering environments into this debate, another approach is required. One way in which geomorphological thinking can contribute is through explicitly recognising spatial variability in the erosion and weathering systems, but through an approach which is able to classify landscapes at a level tractable for analysis. The scale of analysis of the 'landscape scientists' plugs the gap in Figure 3.5.

Geomorphological understanding of landscapes can be codified through linear models such as the sediment cascade (Burt and Allison 2010). The sediment cascade conceptualises systematic changes in sedimentary environment along a continuum from uplands to the coastal zone, or through descriptive analysis of the function of characteristic landsystems (e.g., Evans 2014), which identify distinctive suites of sedimentary environment that characterise particular process regimes, such as the glacial system, cold environment landsystems or peatland landscapes. Key controls on weathering reactions are temperature and the concentration of reactants. The latter is controlled by the nature of the weathered sediment and the rate of removal of weathering products, and so is a function of sediment type, hydrology and rates of erosion. Figure 3.6 describes key controls on rates of weathering. At the range of scales which constitute the missing scale

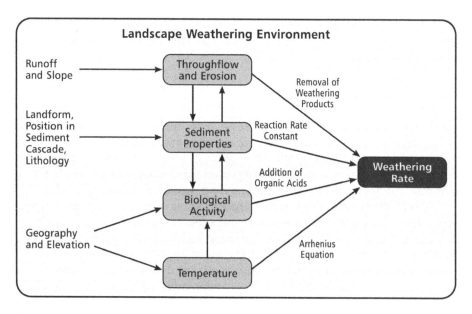

Figure 3.6   Geomorphological controls on weathering rate. Source: Martin Evans.

of analysis in Figure 3.5, these processes can be conceptualised as the landscape weathering environment. Integration of erosion, hydrology and vegetation in this conceptualisation has very strong congruence with the aims of a land system classification, such as the cold environments land system illustrated in Figure 3.7. This incorporates an understanding of the sediment cascade, with classification of characteristic sedimentary environments.

Simple regionalisations of weathering regime could be based on easily mapped environmental parameters such as temperature and rainfall which are important controls on weathering rate. However, such an approach does not take account of the linkages between sedimentary environment and weathering regime, which are implicit in Goudie and Viles' (2012) call for the development of a geomorphically situated understanding of controls on global weathering rates. Developing this understanding through a land systems approach to define weathering landscapes would provide the potential to produce distributed models of global weathering rates, based on mapping of geomorphology, vegetation and climate. Such an approach would make space for understanding of landforms and the connectivity of sediment fluxes in a way that is often lost in global and continental scale analysis of material flux, and provide a basis for meaningful subdivision of the continuum of terrestrial weathering environments. The suggestion by Brantley and Lebedeva (2011) that vertical patterns of regolith chemistry can be interpreted as long-term integrated proxies for weathering environment and their call to

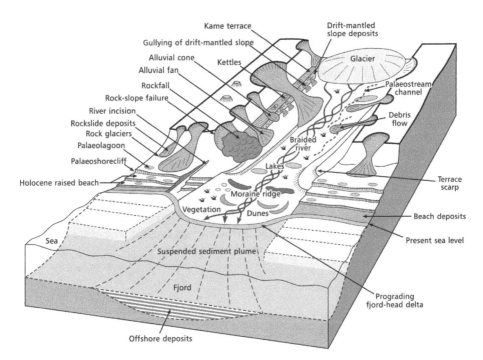

Figure 3.7   A cold environments landsystem. Source: After Tweed et al. 2007. Courtesy of Denis Mercier.

extend this work to a range of geomorphological contexts indicates one promising way forward.

Opening up the black box of continental scale approaches has the potential to develop a more nuanced understanding of the linkages between continental weathering and geological carbon cycling. Considering geological carbon cycling in this framework also entails working at time and space scales closer to those of the fast carbon cycle, raising the possibility that analysis of weathering landscapes could usefully be integrated with analysis of organic carbon fluxes to consider integrated carbon landscapes.

Global Change and the Geological Carbon Cycle

Anthropogenic climate change is understood to be driven by changes in atmospheric greenhouse gases including carbon dioxide and methane (Cubasch et al. 2013). The main driver for this has been the burning of fossil fuels, which leads to a transfer of carbon from the lithosphere to the atmosphere. Rates of CO_2 release to the atmosphere through this mechanism are almost two orders of magnitude greater than natural release from volcanism or weathering of rock-derived organic carbon (7.8 PgC yr^{-1} (fossil fuels) vs. 0.1 Pg C yr^{-1} (volcanism); Cubasch et al. 2013 and 0.04–0.1 PgC yr^{-1} (Rock OC) Hilton and West 2020) Anthropogenic forcing of CO_2 release is therefore equivalent to a massive acceleration of one term of the geological carbon cycle. As a consequence of the long timescales over which the geological carbon cycle operates, and the small annual fluxes (relative to the fast carbon cycle), the impact of this acceleration is primarily experienced through feedback to the fast carbon cycle. It has been argued in this chapter that geomorphological thinking can make a contribution to the long-term understanding of the earth system that is being developed through ongoing research on the geological carbon cycle, and that the added value of this approach comes through considering the problem at reduced time and space scales, to provide geographical granularity to understanding of the system. However, the extensive geomorphological expertise in assessing material fluxes across changing landscapes and at a wide range of timescales also maps directly onto analysis of processes which drive the fast carbon cycle, and so has the potential to develop understanding of the way in which anthropogenic acceleration of the geological carbon cycle impacts on terrestrial carbon cycling. These are the ideas developed in subsequent chapters of this book.

Chapter Four
Geomorphological Theory and Practice
Material Fluxes in the Terrestrial Carbon Cycle

Introduction

Chapters 2 and 3 have outlined key processes controlling the carbon cycle, including biological and sedimentological controls on the flux of carbon through the terrestrial system. Alongside the many technical and measurement challenges which face carbon science when characterising terrestrial systems, a series of conceptual challenges also arise, which derive from the need to integrate diverse mechanisms operating at a wide range of temporal and spatial scales. The premise of this book is that a complete analysis of the system requires interdisciplinary understanding bringing together the geomorphologist's expertise in sedimentary dynamics, with the biologist's knowledge of carbon cycling through living systems. Biological and ecological concepts such as the river continuum concept (Vannote et al. 1980) have been widely applied to understand landscape-scale carbon flux. However, integration of conceptual approaches to understanding sedimentary dynamics in the landscape that are derived from the geomorphological literature has been limited. This chapter reviews a range of influential geomorphological theory, which has the potential to be incorporated into a more complete understanding of the terrestrial carbon cycle at temporal and spatial scales that are relevant to both the biological and sedimentary system.

Geomorphology and the Carbon Cycle, First Edition. Martin Evans.
© 2022 Royal Geographical Society (with the Institute of British Geographers). Published 2022 by John Wiley & Sons Ltd.

Catchments as Integrating Concepts – The Sediment Cascade

Cole et al. (2007) argued that the role of freshwaters in the global carbon cycle had been significantly underestimated and that through their integration of catchment function, drainage networks were a key component in 'plumbing the global carbon cycle'. The focus of this conceptualisation however is strongly on carbon fluxes within the fluvial system. However, the flux of organic matter in the solid state through the landscape system is only partially understood by a focus on the fluvial system, since key elements of hillslope processes and their linkage to the fluvial system are not fully described.

In this book the concept of the sediment cascade (Burt and Allison 2010) is taken as a central organising principle to describe erosion and deposition of sediment and organic matter in the landscape system, and the subsequent flux of terrestrially derived carbon from continents to oceans. The concept of the sediment cascade has evolved from the importance of systems theory in geomorphological understanding (Chorley and Kennedy 1971; Strahler 1980). By conceptualising the landscape as a series of geomorphological subsystems (defined by landform or by process regime), through which energy and matter flow discontinuously over a variety of timescales (Figure 4.1), the idea of the sediment cascade focuses attention on the magnitude of material flux, and the degree to which connectivity

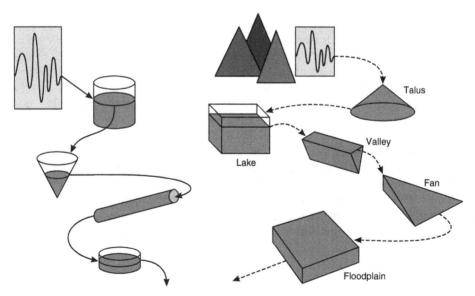

Figure 4.1   Conceptual representation of the sediment cascade (right) characterised by intermittent sediment transfers in contrast to continuous flows of water (left) through the system. Source: After Davies and Korup 2010. Reproduced with permission of Wiley.

between sub-systems is a key control on sediment flux, from source (headwaters) to sink (oceanic deposition).

The idea of the sediment cascade as a system defined by zones of sediment storage linked by the processes of sediment erosion, transport and deposition, is central to most of the ways in which geomorphologists have understood terrestrial sediment flux. Recent geomorphological work has built on the history of understanding of the sediment cascade, and the connectivity of sediment systems, to begin to quantify the degree of coupling within a sediment system. Heckmann and Schwanghart (2013) have developed graph theory approaches to quantify connectivity in alpine sediment systems. They modelled process domains based on terrain analysis using digital elevation data and demonstrated for a particular case study site, that 95% of the catchment area might be regarded as structurally disconnected. They also developed the concept of the effective catchment area, which defines the source area for sediment delivery to a downstream point, as a fraction of the hydrologically defined catchment area.

Such an approach has potential for modelling landscape-scale erosional carbon fluxes, since it allows a spatially dynamic sediment delivery ratio to be determined for a landscape from elevation data. To fully understand landscape carbon flux, the static model of structural connectivity requires coupling to a dynamic model to allow for the evolution of connectivity over time in response to landscape change. The approach of Torres et al. (2017), which uses a meander migration model to produce dynamic hillslope coupling when modelling POC flux, represents a useful way forward here.

The Sediment Budget Approach

The sediment budget (Dietrich et al. 1982) has become a core conceptual and methodological approach in geomorphology (Slaymaker 2009). Sediment budget approaches identify and quantify the main stores and fluxes of sediment within a drainage basin, and are a key methodological intersection between understanding landscape form, and the role of earth surface processes in driving biogeochemical cycling (Slaymaker and Spencer 1998). Producing a quantified sediment budget requires geomorphological expertise to identify the key elements of the budget, in order to construct a conceptual sediment budget model (Figure 4.2). Quantification of a sediment budget requires deployment of a wide range of geomorphological techniques, which span both modern process monitoring and historical geomorphological approaches, based on geochronology and environmental reconstruction. Increasingly valuable in sediment budget studies is the ability to quantify volumes of storage in depositional landforms using digital elevation data, and also to estimate rates of change from repeat DEM surveys (e.g., Ham and Church 2000).

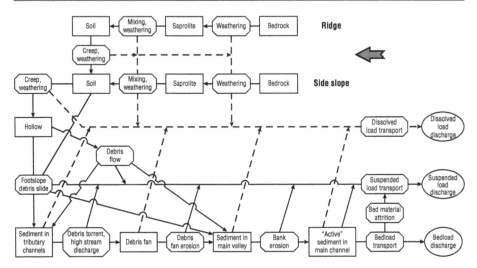

Figure 4.2   Conceptual sediment budget model of the Rock Creek basin, Oregon. Rectangles denote storage, octagons transfers and circles outputs. Solid lines denote sediment transfer and dashed lines denote solute flux. Source: After Dietrich and Dunne 1978. Reproduced with permission of Schweizerbart (http://www.borntraeger-cramer.de/journals/zfg).

Although the sediment budget was developed primarily as a method to assess sediment flux and landform interactions, the formulation is directly analogous to a systems approach to biogeochemical cycling. Galy et al. (2015) have argued sedimentary processes are the dominant control on particulate carbon flux through the sediment cascade, so that the value of understanding particulate carbon fluxes in the context of a sediment budget is significant. At the simplest level, a knowledge of the organic matter content of key sediment types within the sediment budget allows direct translation of the sediment budget into a particulate carbon budget.

Integral to useful application of the sediment budget concept is selection of an appropriate timescale for analysis. The classic sediment budget studies of Trimble (1981) (discussed in Chapter 6) developed sediment budgets for a range of time periods, which demonstrated significant change in storage over space and time, whilst net sediment output to the main stem river remained relatively constant.

The sediment budget as a conceptual framework has become an important organising concept for geomorphology (Slaymaker 2003), and is firmly established as a key element of geomorphological analysis relevant to environmental management (Walling and Collins 2008), and to the analysis of a wide range of geomorphological contexts (e.g., Aagaard and Sørensen 2013; Evans et al. 2006; Rainato et al. 2017; Warburton 1990). The influential and widely cited formulation of the sediment budget concept by Dietrich et al. (1982) presented three main elements to create a sediment budget: 1) recognising sediment transport processes; 2) identifying linkages between sediment stores; and 3) quantifying sediment stores. Arguably, the most influential component of this work has been the conceptual exercise of defining the sediment system in terms of its key fluxes and stores as a basis for further analysis (Figure 4.2; Dietrich and Dunne 1978). Most studies also quantify the main elements of the sediment budget (e.g., Figure 4.3). However, a major challenge in operationalising the sediment budget approach has been assessing the time dependency of the budget. In particular, approaches to incorporating a wide range of processes with divergent magnitude and frequency characteristics are critical to developing representative sediment budgets.

The challenge of integrating measurements taken at a variety of timescales into a sediment budget was acknowledged in the original formulation of the concept by Dietrich et al. (1982), who recognised that elevated sediment flux after hillslope disturbance occurs at timescales of hundreds to thousands of years. The sediment budget derived from short-term monitoring is therefore a snapshot of the system. A longer-term perspective requires analysis of sediment flux at timescales compatible with sediment storage times within the system. Dietrich et al.

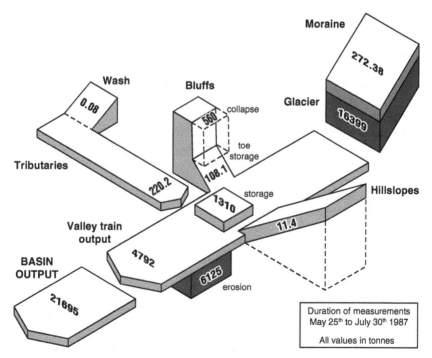

Figure 4.3   Example of a quantified sediment budget for the proglacial fluvial system at Bas glacier, d'Arolla in Switzerland. Source: After Warburton 1990. Reproduced with permission of Taylor & Francis.

developed a description of sediment storage based on reservoir theory, so that the average age of material in the store is derived from an age distribution function, and the age of sediment leaving the store is subsequently used to derive an average transit time of material through the store. The mathematical formulation developed by Dietrich et al. allows estimation of transit times for multiple stores within the system. In the study of age distributions and transit times of organic matter within sedimentary stores, radiocarbon dating is a powerful method which has begun to be exploited for this purpose (e.g., Alderson et al. 2019a).

The potential application of the sediment budget approach to an understanding of carbon cycling is clear. Where organic matter is stored in sedimentary features, the net gaseous carbon losses will be a function of storage time, and transit times of sediment through the system will determine relevant timescales for the averaging of downstream organic carbon flux.

Connectivity

The sediment budget approach emphasises identification of the linkages between sediment stores, and the magnitude of fluxes between those stores. Another key concept, which geomorphologists have developed for the analysis of this element of the sediment system is that of connectivity. Characterising connectivity between elements of environmental systems is fundamental to understanding fluxes of material and energy, and the importance of connectivity has been recognised in a range of disciplines, including hydrology (Bracken and Croke 2007) and ecology (Baguette et al. 2013; Merriam 1984), as well as in geomorphological understanding of the sediment cascade. High transient form ratios in dynamic landscapes imply that depositional landforms are rapidly eroded and reformed, so that the residence time of sediment and organic matter is reduced, than in more stable landscapes with a lower ratio. One way in which geomorphologists have understood this element of landscape change, and conceptualised the controls on the flux of sediment through the sediment cascade, is through consideration of connectivity.

Because connectivity is a widely used concept, it has been used in a variety of contexts and Wainwright et al. (2011) have emphasised the importance of precision of usage if the concept is to aid understanding of environmental flows. They define two types of connectivity as follows:

1) *Structural connectivity*: describes the physical connection or proximity of two elements of the system. For example, a river channel would be considered as structurally connected to its floodplain.
2) *Functional connectivity*: describes the ways in which structural linkages impact on system function, for example the impact of intermittent hydrological linkage between a floodplain and river on nutrient cycling in the channel.

Wainwright et al. emphasise that whilst the division simplifies operationalisation of the concept (e.g., structural connectivity may be quantified through indices of contiguity; Wu and Murray 2008), a complete understanding of material flux at landscape scale must consider interactions of structural and functional connectivity. Only where systems exhibit both elements of connectivity is material flux maximised. In a sediment system, the conjunction of structural and functional connectivity may be episodic in time, driven for example by the generation of overland flow.

More specifically, geomorphologists have focussed on the concept of sediment connectivity, which describes the processes that control the movement of sediment between stores (depositional landforms) in the sediment system, and the rate and time-course of the movement of those sediments. Bracken et al. (2015) describe sediment connectivity as 'dependent not on individual processes … but also on emergent characteristics of sediment deposition and sediment residence times', p. 178.

Many of the earlier approaches to conceptualising connectivity such as considerations of sediment system coupling, or a focus on the sediment delivery ratio as a manifestation of system connectedness, have proved difficult to generalise. In particular, determination of appropriate temporal and spatial scales to assess connectivity has been challenging.

Bracken et al. (2015) have attempted to address these issues by proposing a framework for understanding connectivity that is based on sediment travel distance and entrainment, with connectivity defined as 'the connected transfer of sediment from source to sink in a catchment, and movement of sediment between different zones within a catchment…' p. 177. This approach emphasises the typical travel distances of sediment particles within the system as a metric of connectivity.

This is a critical distinction in the context of the application of the concept to analysis of terrestrial carbon cycling. A proportion of sedimentary carbon that is adsorbed to mineral particles, or enclosed in mineral aggregates, will have characteristic travel distances controlled by the characteristics of the mineral substrate. However, fresh litter and unbound organic matter – the more labile components of sedimentary carbon – have lower density, and may be selectively entrained in erosion events and have longer characteristic travel distances. In this context, the functional connectivity of the carbon system may differ from that of the wider sediment system.

An alternative approach to developing a conceptualisation of connectivity based on particle transport is to consider the role of landforms and landscape configuration as structural controls on sediment flux. Fryirs (2013) argued that connectivity or disconnectivity between hillslopes and channels is a function of fluvial form. Narrow valleys have stronger connectivity and sediment conveyance because of closer linkage of slopes and channels, whereas broader valley systems with a range of depositional landforms such as fans, terraces and broad floodplains reduce connectivity in the sediment cascade. The value of this approach is that it associates connectivity with particular suites of landforms and landform configurations. In the context of a carbon landsystem, this means that there is potential to define characteristic ranges of connectivity associated with distinct landscape types.

Arguably the critical linkage in geomorphological systems is between hillslopes and channels. Here connectivity varies both in space (as a function of topography and runoff) and in time, where the role of high magnitude low frequency mass wasting events can be critical to sediment delivery. Croissant et al. (2019) explicitly model connectivity across this boundary in steep seismically active landscapes in the southern Alps showing that sediment evacuation is controlled by hillslope connectivity and by rates of fluvial reworking. Model results indicate the potential that limited connectivity can lead to sediment accumulation in these headwater systems. Here limited connectivity creates depositional sedimentary environments where carbon may be stored or mineralized. Effectively, an understanding of connectivity in the sediment cascade provides a physical template for the understanding of carbon flux.

Integrating Connectivity and Residence Time in Quantitative Sediment Budget Studies

A major challenge in moving beyond description in sediment budget studies, has been the difficulty in how to treat the scale dependency that is inherent in standard formulations of the sediment budget. Whilst the theoretical approaches described above provide an important context, they represent a series of interlinked concepts, which have proved hard to operationalise, perhaps leading the geomorphologist's uneasy relationship with theory. Recent work by Hoffmann (2015) has developed a new approach to considering appropriate timescales in sediment budgets, which integrates a range of classical geomorphological understanding into a new approach to analysis of the sediment cascade. Hoffmann has developed an approach to sediment budgeting which moves beyond simple proportional characterisation of erosion and deposition of sediment stores (effectively the sediment delivery concept), and focuses on sediment transit times through the system.

Hoffmann's approach is a mathematical framework that integrates reservoir theory into sediment budget approaches. In this method, storage within the sediment cascade (depositional landforms) is the starting point for analysis. Sediment flux within the system is controlled by connectivity between depositional landforms, which also determines the residence time of material in the landform unit, and consequently the age of that material. Hoffmann demonstrates that a focus on the proportion of material lost from a store (as a measure of connectivity) which he terms the 'specific rate', is less scale dependent than using traditional measures such as the sediment delivery ratio. The challenge of scale is not completely removed, since as Hoffmann argues, the specific rate may reduce in time as landforms stabilise or are progressively removed in space from the locus of erosion. However, the framework proposed by

Hoffmann is potentially a valuable step forward, and offers the potential to integrate historical and geochronological approaches with contemporary process studies, to develop quantitative understanding of the sediment cascade.

In an application of the approach to a small catchment in Northern Bavaria, Hoffmann demonstrates that the system is in disequilibrium in response to agriculturally enhanced soil erosion starting 5000 years ago. Critically, the approach is able to estimate residence times for key elements of the sediment cascade, indicating that hillslope sediments have residence times five times those recorded in the floodplain system.

Landscape Position

The sediment cascade is a key conceptual model for the understanding of material fluxes through the landscape system as discussed above. However, it is important to recognise that it is a partial model. The key caveat is that the sediment cascade is primarily a description of gravity driven processes. This is one of the key strengths of the conceptual model, since a necessary consequence is that flux through the system is directional from source to sink. Hillslope processes and fluvial processes move material downslope, so that it is reasonable to consider the catchment as a fundamental unit of geomorphology. However, critically the sediment cascade does not deal effectively with processes that are not driven by flows of water or gravitationally driven hillslope processes.

Aeolian processes, which may move mass and energy between drainage basins are not well captured by this conceptualisation. It has been argued that atmospheric dust deposition linked to Aeolian erosion is an important driver of oceanic carbon sequestration, so that there are potential negative feedbacks between stabilisation of continental eroding sites and oceanic carbon storage (Ridgwell 2002). Consequently, this is an important area for integration into carbon modelling but has not been widely addressed.

However, the sediment cascade model does capture the principle carbon fluxes that are considered in the current IPCC methodology (Cubasch et al. 2013). The role of topography as a key driver in the sediment cascade establishes the potential for upscaling of understanding of carbon landscape models to global scales. The easy availability of high resolution digital elevation data at global scales, provides potential methodologies for mapping carbon landsystems as a basis for upscaling of local process understanding. Mapping of landform elements and terrain position from digital elevation modelling is now routinely applied (Schmidt and Hewitt 2004; Skidmore 1990) and a range of approaches based on relative topographic position have been developed, which have been able to model slope position and landform character (De Reu et al. 2013). These approaches have further been used to model soil carbon, but importantly only in the upper 30 cm of the soil profile (Thompson and Kolka 2005).

The importance of slope position as a key descriptor integrating elements of the sedimentary carbon system, has been recognised in research analysing lateral transport and deposition of carbon within the sediment system. In particular, the work of Berhe et al. (2012) (see Chapter 6) has demonstrated distinct processes and timescales of carbon turnover in top, mid- and footslopes.

Persistence and Landscape Sensitivity

Periods of landscape stability in the order of decades typically produce climax vegetation stands and stable rates of carbon sequestration in soils and sediments. However, understanding the terrestrial carbon flux at longer timescales requires an understanding of the typical timescales of persistence of landforms. Periods of carbon storage cannot exceed the lifespan of the depositional setting where the carbon is being stored. The concept of landform stability and persistence is useful here, since analysis of landform persistence defines the available time for carbon turnover (by biological processes), and the possible timescales for carbon terrestrial sequestration. Analysis of patterns of erosion and deposition at longer timescales are a core element of historical geomorphology (e.g., Macklin and Lewin 2008), which provides the empirical basis to assess relevant timescales. Geomorphologists have also grappled with appropriate conceptual frameworks to understand the response of landscape systems to multiple drivers of change operating at divergent timescales, and these ideas have relevance to attempts to integrate biological and sedimentological elements of the carbon cycle.

In the classical geomorphological literature, Brunsden and Thornes (1979) defined the question of landform persistence in the landscape through developing the Transient Form Ratio (Figure 4.4), which is $TF_r = R/D$, where R is the mean relaxation time for the landscape, and D is the recurrence interval of events which perturb the system (effectively major erosion or deposition events). A landscape where recurrence intervals exceed relaxation times will have a low proportion of transient forms and a high degree of landform stability, controlled by the processes of landscape disturbance. Under these conditions, rapid reworking of sediment and high degrees of connectivity through the sediment cascade mean that disturbance events and associated processes of sediment transformation that occur in the freshwater system are likely to be an important control on carbon cycling. A higher transient form ratio implies greater stability of landforms, and would favour biological controls on NPP, and carbon mineralisation in the sedimentary system as the dominant control on carbon flux.

Phillips (1995) discusses the idea of distinguishing transient form ratios for the vegetation and sedimentary systems, arguing that where similar ratios are observed for two components, this can be considered as a distinct landform–vegetation system. However, case study data from the North Carolina coastal plain indicates significant variability in these ratios across landscape features. Phillips

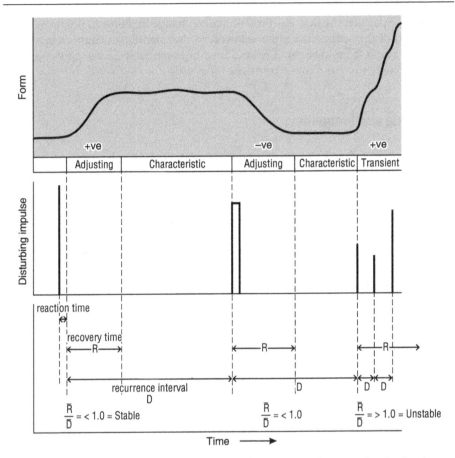

Figure 4.4   Stable and unstable landscape systems defined by the transient form ratio where R is the relaxation time for the landscape and D is the recurrence interval of the disturbing event (e.g., flood or landslide). Source: After Brunsden and Thornes 1979. Reproduced with permission of The Royal Geographical Society.

argues for the need for better estimates of rates of change and relaxation times, to allow more generally applicable conclusions about vegetation–sediment system relations. Nevertheless, it is argued that the transient form ratio, or other similar estimates of landscape stability discussed by Phillips, represent a useful classification of landscape types in the context of understanding carbon cycling through the sediment cascade.

A related concept is geomorphic or landscape sensitivity. Brunsden and Thornes (1979) defined this as the ability of landforms to persist after changes in the process regime. The landscape sensitivity of a sediment system is characterised by the 'potential or likely magnitude of change within a physical system and the ability of that system to resist change' (Thomas and Allison 1993). In a sensitive landscape, the degree of resistance to change inherent in the form (configuration) of the sediment system is lower, so that significant changes in system state may result from lower energy land forming events. Sensitive landscapes are therefore likely to have a lower transient form ratio.

Magnitude and Frequency

Disturbance is a key element of understanding ecosystem change. Reworking of depositional landforms and the associated stores of sediment and soil carbon occurs in response to geomorphological change driven by synoptic forcings such as wind and rain events. The magnitude and frequency of these impulses for change define the temporal patterns of sediment and carbon flux, and also directly impact the assessment of landscape sensitivity and transience described above. Considerations of magnitude and frequency have been central to geomorphological thought since the work of Wolman and Miller (1960). Central to this concept is that the maximum amount of geomorphic work is completed by events of moderate frequency and magnitude (Figure 4.5). Magnitude and frequency concepts in geomorphology are typically used to assess the relative importance of erosional and

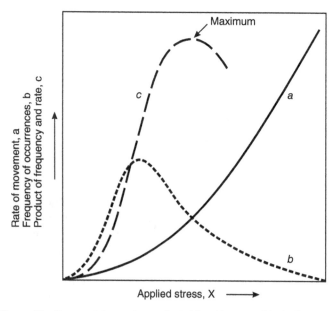

Figure 4.5   The combined impact of changes in magnitude (a) and frequency (b) of sediment transport events on total sediment transport. Source: After Wolman and Miller 1960. Reproduced with permission of University of Chicago Press.

depositional events at given timescales, in order to estimate reasonable long-term rates of change. For example, Reid and Page (2003) combined data on the recurrence and scale of landsliding in New Zealand (derived from sequential aerial photography) with long-term rainfall data, to estimate 100-year average sediment fluxes. This is highly relevant to attempts to derive representative rates of sediment associated carbon flux within catchments. The assessment of the frequency of high magnitude events within the landscape is also an important element of understanding ecosystem disturbance and its impact on primary productivity.

Closely linked to notions of magnitude and frequency are considerations of geomorphic thresholds (Schumm 1979) and equilibrium states. Geomorphic thresholds are points of rapid transition between system states. Schumm distinguished intrinsic thresholds where a change in system state might result from random variation within a system, and extrinsic thresholds where change in an external driver (such as changing climate) triggers a change in the equilibrium state of the system (Figure 4.6). In either case, the magnitude of the disturbing event has to be sufficient to exceed the threshold and trigger a change in system state. Figure 4.6, following the work of Chorley and Kennedy (1971), characterises the nature of the response for systems in stable, metastable or unstable equilibrium conditions. The nature of sedimentary change over time, as conceptualised in Figure 4.6, is highly significant for understanding the ways in which interactions of the sedimentary and biological system control carbon sequestration. Periods of relative landscape stability favour biological control of the surface carbon balance, whereas during periods of rapid change entailing erosion and deposition across the landscape, sedimentary processes of disturbance and burial become significant controls on carbon cycling.

Figure 4.5 can also be examined with reference to thresholds. It assumes that the work done is proportional to the force exerted and the frequency of the event, but where threshold energy levels are required to initiate change, for example the crossing of shear stress thresholds to initiate particle movement, this may not hold. Thresholds of change may also link to the development of connectivity within the landscape. For example, where overland flow is generated in a patchy manner controlled by spatial patterns of vegetation and soil character, significant solute flux and erosion may not be generated until the disjunctive patches join to generate continuous flow across the land surface. Effectively, a threshold is crossed when the structural and functional connectivity of this system become aligned so that material flux across the hillslope is unimpeded.

The Sediment Delivery Ratio

A widely used approach to summarising the connected sediment flux through catchment systems is the sediment delivery ratio (Glymph 1954; Walling 1983). This is typically defined as the ratio of the measured sediment yield at the catchment outlet to total erosion in the catchment. High sediment delivery ratios

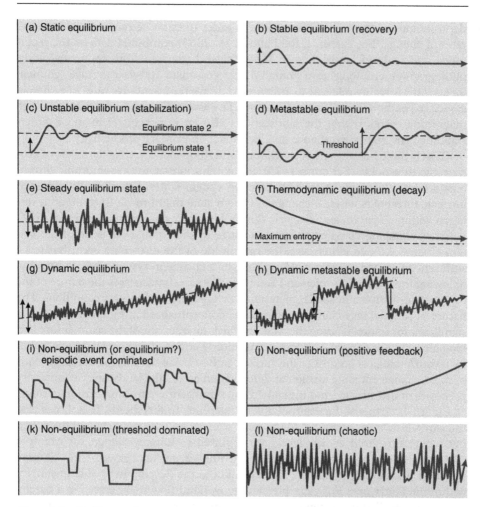

Figure 4.6   Classification of types of geomorphological equilibrium. Source: After Huggett 2003. Reproduced with permission of Taylor & Francis.

imply lower storage within a catchment and vice versa. The SDR is a widely applied concept, which provides a conceptually simple description of the relative importance of erosion and deposition within a catchment. The SDR in this context is a useful descriptor when considering terrestrial carbon cycling, since it encompasses the potential for lateral export of organic sediments and the burial of organic carbon within depositional landforms.

However, the simplicity of a spatially lumped sediment delivery ratio as a descriptor of landscape function has the potential to mask important complexity within the system. Walling (1983) noted the black box nature of the concept, and highlighted a series of challenges relating to the use of SDR as a key descriptor, including a lack of process understanding resulting from both the spatial and temporal lumping inherent in the approach, and the lack of consideration of sediment quality.

An important way in which understanding of sediment delivery ratio has been deployed is in the analysis of the influence of spatial scale on observed sediment yield. Plots of specific sediment yield against catchment area commonly exhibit a declining trend, which is interpreted as a lower sediment delivery ratio in larger catchments due to enhanced storage on lower slopes. Exceptions to this pattern are observed where significant additional sediment sources are present in larger-scale catchments. For example, Church and Slaymaker (1989) demonstrated that a pulse of paraglacial sediment reworked through fluvial systems in British Columbia at Holocene timescales, led to larger specific sediment yields in mid-sized catchments. Similarly, Worrall et al. (2014) have argued that variable trends in UK rivers may be linked to changes in the relative size of headwater sediment sources, proportional to catchment area with increasing catchment size. The SDR as a simple measure does not discriminate between changes in gross erosion and changes in deposition. In the context of carbon sequestration, the magnitude of organic sediment deposition may vary significantly for a given SDR. Worrall et al. (2014) consider that spatial patterns of SDR are inherently flawed as a method of considering changes in catchment sediment flux because of the intrinsic self-correlation of plots of specific sediment yield (t/km^2 a^{-1}) against catchment area (km^2). They argue that interpretation of simple plots of sediment flux against catchment area are more informative, with inflexions in empirical s-curves (Gompertz functions) fitted to the data, indicating potential changes in process regime from erosion, to transport, to deposition.

Because of the conceptual simplicity of the SDR, a range of attempts to model the ratio in space and develop appropriate scalings have been developed, including fractal scalings (Zhang et al. 2015) and modelling of spatial patterns of SDR based on landscape connectivity metrics derived from digital elevation models (Vigiak et al. 2012). De Vente et al. (2007) reviewed applications of the concept, and concluded that attempts to understand spatial variability in SDR could not rely on simple relations to area, but that a model incorporating spatial patterns of land use, climate, lithology and topography was required. Increasingly, the usefulness of the concept has been questioned. Lu et al. (2005) suggest that spatial and temporal variability in sediment source area and in depositional processes, mean that a simple comparison of upland erosion rates with catchment sediment yield may not be informative. They argue that at equilibrium timescales, SDR must approach unity and that SDR measured above or below one results either from difficulty of measurement or from analysis at a timescale shorter than the relaxation time of the system. Parsons et al. (2006) have similarly argued that the basic SDR concept is a fallacy when expressed in units of mass per unit area per unit time, because it is impossible to properly define the sediment contributing areas in a catchment. They argue that analysis of the sediment cascade should focus on spatial and temporal patterns of sediment flux, rather than on an arbritary ratio of upland to lowland flux.

The SDR, despite these issues, is widely calculated and referenced in the geomorphological literature. An uncritical application of the concept to understanding lateral fluxes of sediment associated carbon may be misleading for the reason described above. However, it potentially provides a metric of sediment system state that is useful in the context of landscape carbon flux. An SDR close to unity could be a useful indicator that the timescale at which the calculation is made is one at which the sediment system is in equilibrium. This is an important consideration in comparing sediment associated carbon fluxes over long timescales, with more rapid carbon turnover in the biological system.

Biogeomorphological Theory and Carbon Cycling

Biogeomorphology is the interdisciplinary study of the interactions between ecological processes and geomorphological systems (Naylor et al. 2002; Viles 1988). Biogeomorphological research has highlighted processes and feedbacks through which geomorphological processes impact on the biological function of ecosystems, and ways in which biological influences are integral to land-forming processes (Table 4.1).

Table 4.1   Ecological impacts on geomorphological processes and geomorphological impacts on ecological systems.

	Stabilising	*Destabilising*
Ecological impacts on geomorphological systems	• Vegetation growth reduces erosion • Microphytic crusts reduce erosion • Vegetation growth enhances sediment storage • Large woody debris in fluvial systems decreases flooding and enhances sedimentation • Animal grazing enhances vegetation cover and reduces erosion	• Animal burrowing enhances erosion • Animal disturbance to microphytic crusts enhances erosion • Beaver dams disrupt fluvial systems • Large woody debris in fluvial systems increases flooding • Animal grazing reduces vegetation cover and enhances erosion
Geomorphological impacts on ecological systems	• Weathering produces nutrients for biofilm growth • Sediment deposition provides growth substrate for vegetation • Sediments provide burrowing substrate for animals	• Weathering removes biofilm niche • Sediment deposition buries vegetation • Mass movements destroy burrow systems

Source: After Viles et al. 2008. Reproduced by permission of Wiley

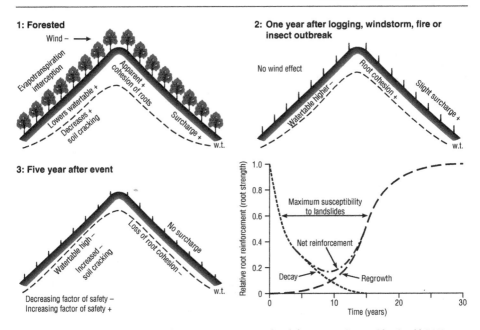

Figure 4.7   Landslide risk has a non-linear response to time after deforestation. Source: After Pawlik 2013. Reproduced with permission of Elsevier.

The conceptualisation of the interdependence of the physical and biological processes however, moves beyond simple linear interdependency (e.g., Figure 4.7). Stallins (2006) argue that the interactions are best viewed through complexity theory, so that geographical variation in biogeomorphological interactions can be attributed to semi-stable assembly states, which may be sensitively dependent on initial conditions and subject to rapid change in state. Stallins argues that 'topology (patterns of stable states) for many biogeomorphic systems may ultimately be constrained by the sediment budget', p. 212. The implication is that where geomorphological change is very fast or very slow, emergent patterns of ecological stability may be subsumed by a dominance of sedimentological, or ecological control on ecosystem pattern. Viles et al. (2008) similarly argue that diversity of ecosystem response to extrinsic drivers of change is contingent on the particular configuration of geomorphological and ecological processes operating at a point in time and space. Such complexity poses a major challenge for integrating biogeomorphological understanding into models of carbon cycling. It suggests that developing understanding of carbon flux through the sediment cascade should not focus on globally applicable empirical relations, but rather should focus on developing a process understanding at the scale of the carbon landsystem.

Viles et al. (2008) identified a range of conceptual models relating extrinsic change in climate to a biogeomorphological response (Figure 4.8). They argue

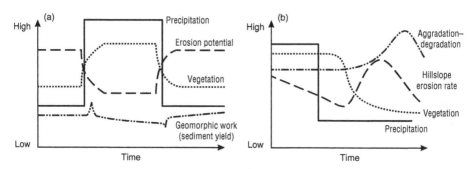

Figure 4.8   (a) Conceptual descriptions of ecosystem response to disturbance showing a relation between erosion and vegetation. (b) Change in carbon storage in a salt marsh system response to precipitation change. Source: After Viles et al. 2008. Reproduced with permission of Wiley.

that further empirical understanding of the timescales over which such responses occur is required in order to generalise such findings. In the context of understanding terrestrial carbon cycling, the impact of these perturbations on ecosystem carbon storage is an additional requirement. Viles et al. present a conceptual plot of this type (Figure 4.8b), describing reduced carbon storage in salt marsh systems eroding due to drought conditions. Deploying existing knowledge of ecosystem response in this manner for carbon landsystems across the sediment cascade, and quantifying the magnitude of such responses, is an important research challenge for biogeomorphology.

Biogeomorphology covers a wide range of scales, but in the literature there is a clear separation of research which focusses on macro-organisms and on patterns of succession, and ecosystem form, function and work, which explores the role of micro-organisms in controlling geomorphological processes, in particular as agents of weathering or bio-protection of earth surface materials. The engagement with the understanding and methods of microbiology, and the recognition that the ecosystem functions of microbial organisms have landscape scale impact, is a unique insight within geomorphology. To date, although the importance of biogeomorpological understanding has been highlighted in work on soil carbon, biogeomorphologists have only brought their insights to bear on carbon cycling in rather limited ways. This is surprising given that in one sense much of the content of this book could be defined as biogeomorphology, considering the interactions of organic carbon in the ecosystem, decomposition of organic carbon by microbial processes, and the transport and transformation of organic carbon through the sediment cascade.

Coombes (2016) describes biogeomorphological research at microbial scales as 'comparatively lacking in output and application', p. 2297. Whilst the paucity of output, particularly outside the field of rock weathering is incontestable, the lack of application is arguably simply a matter of focus. Organic carbon, its flux through the sediment cascade, and its role in the global carbon cycle, is literally the 'stuff' of biogeomorphology.

Biogeomorphology is the disciplinary focus that has the potential to integrate studies of the production, transformation and mobilisation of organic carbon in the sediment cascade. Microbial influence is not confined to the fast carbon cycle. Viles (2012) argues that the influence of microbial life on rock weathering varies with timescale, but that at timescales of 10^4–10^6 years (which are relevant to considerations of the slow carbon cycle), microbial processes play a significant role in rock weathering and consequent CO_2 sequestration.

It is argued here that viewing the landscape through the lens of carbon cycling, provides a unique application for biogeomorphology, at both the microbial and ecosystem scale. Disturbance and ecosystem recovery have been key drivers of research in biogeomorphology. The importance of integration of this understanding of both microbial and ecological controls on net ecosystem exchange, with the analysis of the lateral flux of carbon and sediment through the landscape, is the core premise of this book. Biogeomorphology and biogeomorphologists are positioned at a key interdisciplinary nexus to make a critical contribution to this challenge.

Landsystems

Land systems are conceptual models of characteristic assemblages of forms associated with particular environments. They describe patterns of topology, topography and material properties in the landscape, which have been identified as typical in a given environment, and derive from a defined range of driving processes. The land system concept was first developed in Australia as an approach to integrating spatial patterns of geology, geomorphology, ecology and climatology into landscape classification (Christian and Stewart 1947). The approach has been widely applied in practical soil survey and in landscape ecology (Zonneveld 1989). In geomorphology, the approach has typically been descriptive, identifying the range and spatial patterns of landforms and processes, which are deemed to be characteristic of a given environment or landscape. For example, Figure 4.9a summarises these diagrammatically, with supporting text describing key processes and linkages within the landscape systems. The approach has been applied to a range of environments including peatlands (Evans and Warburton 2007) and steeplands (Hewitt 2006). It has been particularly well developed in glacial geomorphology, where characteristic landform assemblages or land systems have been used to infer previous glacial process regimes (Evans et al. 1999 and Figure 4.9b). Evans (2014) describes land systems as 'recurrent patterns of genetically linked landscape units', p. 4. The land system concept has been variously applied across a range of disciplines and has been expanded to cover anthropogenic drivers of change (Verburg et al. 2013), but the essence

(a)

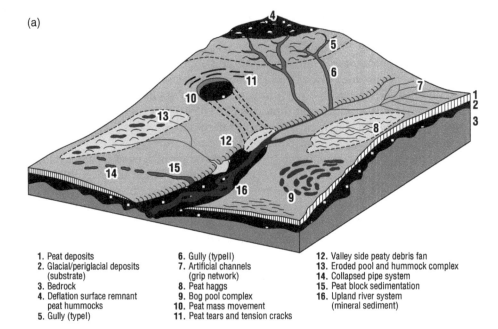

1. Peat deposits
2. Glacial/periglacial deposits (substrate)
3. Bedrock
4. Deflation surface remnant peat hummocks
5. Gully (type I)
6. Gully (type II)
7. Artificial channels (grip network)
8. Peat haggs
9. Bog pool complex
10. Peat mass movement
11. Peat tears and tension cracks
12. Valley side peaty debris fan
13. Eroded pool and hummock complex
14. Collapsed pipe system
15. Peat block sedimentation
16. Upland river system (mineral sediment)

(b)

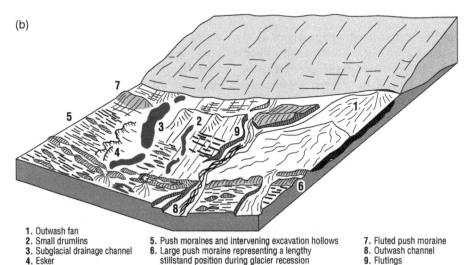

1. Outwash fan
2. Small drumlins
3. Subglacial drainage channel
4. Esker
5. Push moraines and intervening excavation hollows
6. Large push moraine representing a lengthy stillstand position during glacier recession
7. Fluted push moraine
8. Outwash channel
9. Flutings

Figure 4.9 (a) A landsystem model for eroding peatlands. Source: After Evans and Warburton 2007. Reproduced with permission of Wiley. (b) A landsystem model for landscapes of temperate glacier recession. Source: After Evans et al. 1999. Reproduced with permission of Wiley.

of the concept is classification of broad landscape types, with common drivers of change and function. In geomorphology, this has been applied to understand spatial patterns, and to classify elements of the sediment cascade. In the context of the terrestrial carbon cycle, these aspects of the approach are valuable since the definition of 'carbon land systems' may offer a useful conceptual framework to integrate description and analysis of the biotic and sedimentary systems.

Geomorphology, Ecology and Carbon Cycling

The discussion of geomorphological theory above has identified a range of conceptual approaches to understanding the flux of sediment through the sediment cascade. The core of these conceptualisations relates to dealing with the wide range of time and space scales that are required to link a process understanding of geomorphology to a representative analysis of landscape change. Developing an interdisciplinary approach to understanding the terrestrial carbon cycle requires an integration of the process understanding of primary productivity and OM decomposition that ecologists and biogeomorphologists can deliver, with geomorphological understanding of landscape change. Approaches such as the conceptual model of landslide dominated mountain landscapes developed by Restrepo et al. (2009) have demonstrated the importance of such biogeomorphological approaches to the understanding of landscape-scale ecosystem processes, and implicitly therefore to understanding terrestrial carbon cycling.

Figure 4.10 illustrates a simple model output linking at a point carbon storage to ecosystem disturbance. However, it is clear that at the landscape scale, a complete understanding of the carbon cycle includes the integration of processes operating at a wide range of scales to control land surface stability, and to define disturbance regimes in the biological system. Characterisations of land systems and the sediment cascade through the range of geomorphological approaches outlined offers a potential approach to classifying land systems in terms of carbon storage potential. Figure 4.11 incorporates a range of conceptualisations into a plot which characterises a 'carbon turnover space'. It postulates that towards the upper right of the plot, the landscape system is characterised by disturbance and rapid turnover of sediments. In these conditions, carbon turnover may dominate carbon cycling with rapid fixation of carbon at freshly disturbed sites, rapid mineralisation of carbon from disturbed and oxic sediments, and lateral export of large quantities of particulate carbon through a well-connected sediment system. In contrast, towards the lower left of the plot, more stable and less sensitive landscapes, perhaps characterised by lower slopes and larger catchment areas, offer the potential for longer-term storage of organic matter in depositional landforms. Figure 4.11 hypothesises that poorly connected landscapes with lower slopes, characteristic of lowland areas, tend to have higher rates of carbon sequestration,

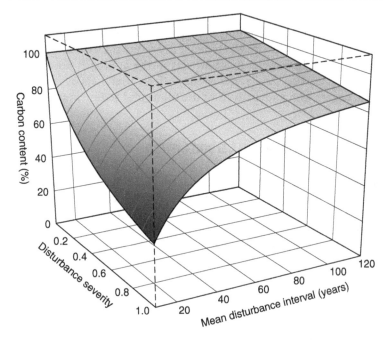

Figure 4.10   General modelling results for ecosystem carbon storage and disturbance. Source: After: Weng et al. 2012. Reproduced with permission of Wiley.

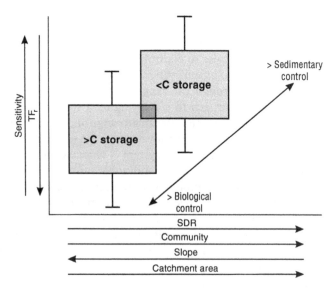

Figure 4.11   A conceptual diagram of a 'Carbon Turnover Space'. Towards the upper right of the plot carbon turnover rates are high in well-connected landscapes, whereas in the lower left of the plot storage dominates more stable systems. Source: Martin Evans.

and are systems where infrequent landscape disturbance means that biological production and processing of organic carbon dominates carbon cycling. In contrast, better connected systems with higher rates of disturbance lead to significant lateral and vertical carbon fluxes influenced by geomorphological processes. In these systems, carbon losses may be high, but high rates of carbon fixation from early successional vegetation, may under some conditions, offset these losses. These systems are therefore sites of rapid carbon turnover, but may function as carbon sources or sinks.

One of the challenges of upscaling process understanding of carbon and sediment interactions to the landscape scale is the choice between continuous and domain specific parameterisations. Approaches, such as the land system model or the use of slope position to characterise carbon sequestration, rely on classification of the carbon landscape and may be better able to capture the complexity of the system, whereas derivation of continuous functions significantly simplifies modelling approaches, but may not have universal applicability in complex systems.

Geomorphological characteristics of landscape systems, based on rates of landscape change to define carbon land systems, require an integration of historical and process approaches to geomorphology, and offer the potential to further integrate understanding of ecosystem responses to disturbance derived from biological and biogeomorphological understanding. In the following chapters, the nature of controls on terrestrial carbon cycling along the sediment cascade is explored, with the aim of integrating biological and geomorphological understanding for a series of carbon land systems from headwaters to coastal systems.

Part II
Geomorphology and Carbon Cycling Across the Sediment Cascade

Chapter Five
Carbon Cycling in Headwater Catchments

Introduction

Headwaters typically comprise 60–80% of the total length of river networks (Benda et al. 2005). The river continuum concept (Vannote et al. 1980) identifies specific roles that headwater streams play within the freshwater terrestrial carbon cycle. Specifically, the RCC suggests that headwater streams are dominated by diverse allochthonous organic matter (OM) derived from catchment vegetation, and that large quantities of coarse particulate organic matter (POM) are processed by shredder organisms, with the resulting POM flowing as an important OM source to downstream waters. Whilst this conceptualisation recognises distinctive patterns of aquatic carbon cycling in headwaters, it is a black box in terms of delivery of OM from the catchment, despite this being the dominant component of OM in these systems. OM fluxes from the hillslopes of the headwater catchments are controlled by geomorphological and hydrological processes operating across the hillslope system (Figure 5.1), and the particular characteristics of headwater systems have been well studied in this regard (e.g., Benda et al. 2004; Gomi et al. 2002; Sidle et al. 2000). Increasingly. the requirement to integrate biological, geomorphological and hydrological understanding to properly understand material flux from headwaters has been recognised (Battin et al. 2008; Gomi et al. 2002).

This chapter aims to review the key characteristics of headwater systems, and explore the implications of this knowledge for developing a more process-based understanding of the controls on flux of OM from headwater systems, and the role of headwaters in terrestrial carbon cycling.

Geomorphology and the Carbon Cycle, First Edition. Martin Evans.
© 2022 Royal Geographical Society (with the Institute of British Geographers). Published 2022 by John Wiley & Sons Ltd.

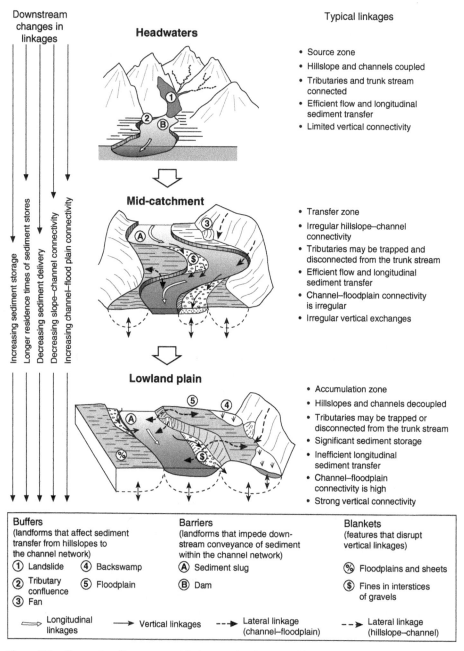

Figure 5.1 Changes in sediment connectivity between headwaters and lowland catchments. Source: After Fryirs 2013. Reproduced with permission of Wiley.

Characteristics of Headwater Systems and Implications for Carbon Cycling

Bishop et al. (2008) describe headwater systems as 'aqua incognita', arguing that headwater systems are numerous, diverse and significantly understudied. The diversity of headwater systems stems from their ubiquity; headwater systems span all biomes, all lithologies and a wide range of topographies. Nevertheless, there are certain characteristics of low-order stream systems which are common across many catchments. Headwater systems are typically steeper and at higher elevation than lowland catchments. Steep slopes, confined valleys and close proximity of channels to sediment sources lead to more effective coupling of material flux between channel and hillslope systems (Benda et al. 2005). This is particularly the case in headwater systems that have been shaped by fluvial and hillslope processes. Over 40% of the global POC flux is derived from land surfaces with slopes of over 10%, which represent only 16% of global land area (Hilton 2017). Where extensive glacial scour has occurred in areas of alpine glaciation, areas of low relief such as cirque basins and hanging valleys may produce local areas of disconnection (e.g., Cavalli et al. 2013; Evans 1997). Small, steep catchments have typically flashy hydrological regimes, and on steep slopes, sediment connectivity between the slope and channel is typically driven by mass movement (Benda et al. 2005). High slope–channel connectivity impacts not only the transfer of sediment and organic matter, but can also modify the microbial communities, which are a key control on the cycling of organic carbon in soil and sediment systems. Besemer et al. (2013) describe headwater systems as 'critical reservoirs for microbial diversity in fluvial networks', p. 5, arguing that headwater systems have rich biodiversity of benthic biofilms, which they explain by recruitment of terrestrial microbes from close slope channel linkages. Crump et al. (2012) show similar results from arctic tundra headwaters, particularly for bacteria and archaea.

The frequency of mass movement in steep headwater systems mean that these are likely to be areas of high disturbance, characterised by relatively immature soils formed on flood or landslide deposits. More generally, low rates of soil formation and relatively short timescales since glaciation,mean that upland soils are characteristically thin and some may be weakly organic. In some cases, litter accumulation can be close to zero. Lipson et al. (2000) studied alpine soils in Colorado and demonstrated that microbial biomass falls after spring snowmelt because the system becomes carbon limited, as all the litter from the previous season has been metabolised.

Headwaters are typically at higher elevation and so are cooler than downstream areas, so that snow and ice processes may be a significant element of catchment function. Changes in temperature with elevation are also likely to directly influence biological processes of carbon fixation and decomposition.

Soil temperature is typically a significant control on rates of respiration (Carey et al. 2016), so that rates of decomposition of soil organic matter are likely to be reduced in higher elevation headwater systems, in relation to places further downstream.

Similarly, primary productivity is also to a degree temperature dependent, although the relation is complex and may be confounded by moisture or nutrient limitations. However, in forest systems, where productivity is often closely monitored for commercial reasons, it is commonly observed that primary productivity declines with elevation (e.g., Girardin et al. 2010; Luo et al. 2004; Raich et al. 1997). However, declining productivity with elevation is not a linear process; where significant thresholds are crossed which impact vegetation type, reductions can be rapid. Perhaps the best example of this is the treeline, which is commonly approximated by the 10 degree Mean July temperature isotherm, or more precisely by mean annual soil temperatures close to 7°C (Körner and Paulsen 2004). A step change in vegetation type across the treeline means changes in productivity are non-linear. Because NPP is commonly correlated with soil respiration, this also implies a more complex relationship with temperature, where elevational gradients cross major ecotonal boundaries. Productivity can also be impacted by topography. In the Sierra Nevada, Mildowski et al. (2015) demonstrated reduced forest biomass on steeper slopes due to higher erosion rates generating thin soils and lower moisture and nutrient retention.

Although temperature effects are likely to lead to somewhat reduced rates of carbon exchange in headwater catchments, the more distinctive features of headwater carbon cycling are likely to relate to the characteristic patterns of slope channel linkage and transfer of organic matter to the stream system. Since headwaters are net exporters of largely allochthonous organic matter, the delivery of material from slopes to channels is a first-order control on the rates of this process.

Figure 5.2 and Table 5.1 after Gomi et al. (2002) compare physical and biological characteristics of headwater catchments with those downstream. Table 5.1 describes the situation typical of forested steepland headwaters, a globally widespread type, and represents a useful attempt at an integrated description of steepland headwaters. However, inevitably because of the diversity of headwater systems, this is not a complete description. For example, it does not assess impacts of snow and ice on headwater systems, which represent a major upland cover type. Glaciers and ice sheets cover 16 million km^2 of land surface; excluding Antarctica and Greenland, there are circa 500 000 km^2 of upland glaciation (Benn and Evans 2014). In some parts of the world, upland peats are also an important component of upland land cover, with characteristic implications for the functioning of headwater catchments. Peatlands cover circa 4 million km^2 of the earth's surface (Xu et al. 2018), a significant proportion of which represent headwater drainage areas. The following sections consider forested headwaters, glaciated systems and peatland headwaters in more detail.

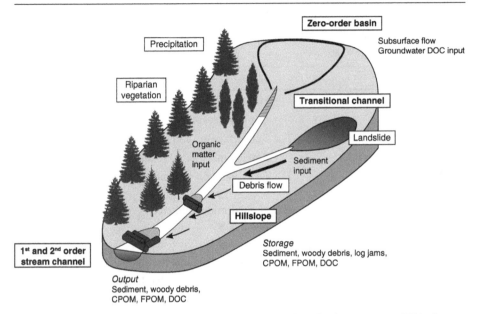

Figure 5.2   Characteristic patterns of sediment and organic matter flux in headwater systems. CPOM is Coarse Particulate Organic Matter, FPOM is Fine Particulate Organic Matter and DOC is Dissolved Organic Carbon. Source: After Gomi et al. 2002. Reproduced with permission of Oxford University Press.

Carbon Cycling in Steepland Headwaters

The ubiquity of landsliding in steepland forests means that the role of mass movement in carbon cycling has been well studied (e.g., Clark et al. 2016; Hilton et al. 2011b; Walker and Shiels 2008). Landslides and debris flows are the dominant mechanisms linking hillslope and channel sediment systems in steeplands, and so play an important role in the terrestrial carbon cycle. Hilton et al. (2011b) estimated that in the southern Alps of New Zealand, landslides transported on average 7.6 ± 2.9 tC km^{-2} a^{-1}, of which 30% is exported to river channels. They also showed that fluvial carbon export in stream systems significantly exceeded this value, which they ascribed to a flux from shallow surface erosion not included in the landslide flux calculations. Hilton et al. demonstrated that in this landscape the high frequency of landsliding means that 0.3% of the land area was subject to landsliding per decade. This high rate of disturbance means that an unusually high proportion of the landscape is in early successional phases, where NPP is higher than in mature forest. A secondary effect of the landsliding is therefore elevated rates of carbon fixation in headwater catchments.

Similar results have been reported for a steepland grassland sites in New Zealand, where Page et al. (2004) showed that 43% of landslide material from steepland

Table 5.1 Interactions of geomorphological, hydrological and biological controls on carbon flux in headwater systems. Based on analysis of forested systems in the northwest of north America (after Gomi et al. 2002). Reproduced by permission of OUP

Process	Headwater Characteristics
Hydrology	High rain/snow
Precipitation	Canopy Closure
Heat Dynamics	Subsurface/groundwater flow
Flow Generation	Smaller more variable discharge
Hyporheic Zone	Dependent on soil and rock flow pathways
Stream Chemistry	
Geomorphology	High altitudes and steep gradients/con-
Morphology	fined valleys
Main Sediment Movement	Episodic Mass Movements
Channel Reach Type	Colluvial, Cascade, Step Pool, Bedrock
Roughness Element	Woody Debris, boulder, bedform (steps)
Biology	Allochthonous and lateral (hillslope)
Energy Input	CPOM > FPOM & DOC from groundwater and
Organic Matter	soil water
Nutrient Source	Groundwater and riparian vegetation
Dominant Functional Group	Shredders
Disturbance	Landslides/Debris flows/Drought

pastoral sites was delivered to water courses yielding circa 50 tC km^{-2} a^{-1}. A further 44 tC km^{-2} a^{-1} was derived from sheetwash, and around 61 tC km^{-2} a^{-1} was fixed through enhanced NPP on recent (<100 years) landslide sites.

Landsliding does not only act as a physical mechanism for the export of terrestrial carbon to the fluvial system. It also plays a significant role in terrestrial carbon cycling through the impact that landslide disturbance has on ecosystems. Turnover of the terrestrial carbon store is largely controlled by microbial decomposition of organic matter. Whilst there have been some studies of microbial succession in response to landsliding, understanding of process links between landsliding, microbial community dynamics and release of terrestrial carbon to the atmosphere is limited.

The impact of landslide disturbance on microbial communities in forested sites has been assessed in a variety of contexts. Myers et al. (2001) argued that variability of microbial communities in upland forest systems is strongly linked to changes in vegetation cover and soil type, which implies that in disturbed upland systems, significant microbial diversity associated with spatial and temporal patterns of landsliding should be expected. Li et al. (2005) measured lower microbial biomass beneath landslide disturbed forest in Puerto Rico, which was correlated with reduced soil carbon content. Landslide impacted sites in Indian tropical forest have been shown to demonstrate reductions in microbial biomass as a proportion of total SOM for four years post-impact (Arunachalam and Upadhyaya 2005). Similarly, DeGrood et al. (2005) demonstrated reduced microbial biomass on road cuts and landslide soils, with recovery to undisturbed levels once sites had re-vegetated. A study of landslide chronosequences at forested sites in Nepal showed that the timescale for restoration of microbial biomass was circa 90 years, whilst complete recovery of soil organic carbon content took around 150–200 years (Singh et al. 2001). It is likely that the enhanced NPP observed on landslide sites, which is in part due to higher rates of carbon fixation in early successional communities, is also related to the time lag between vegetation succession and full recovery of the soil microbial community, which limits decomposition of OC in the early phases of recovery. However, there is limited data on the full carbon budget of these sites associated with suitable microbial data to determine the process basis of carbon dynamics in the aftermath of landslide events. In particular, there is no direct data of the impact of landsliding on gas flux to the atmosphere from landslide sites, which is a significant research gap.

Total carbon flux associated with mass wasting in tropical steeplands has been reviewed by Ramos Scharrón et al. (2012), with carbon fluxes ranging from 3–39 tC km^{-2} a^{-1}. Forest NPP is typically on the order of hundreds of tC km^{-2} a^{-1}, so that the losses from a single sediment transfer process are substantial. In steepland landscapes, the processes transferring organic matter to stream systems are relatively high energy; the role of landsliding has been discussed and it has also been observed that high POC concentrations in forested headwater systems coincide with the highest flows (Hovius et al. 2011). This is consistent with the observations of Turowski et al. (2016) that coarse POC is the dominant component of organic carbon export from a subalpine system in the Swiss Alps. The dominance of coarse particulate organic matter (CPOM) is predicted by the river continuum concept, but methodologically this observation is important, since coarse material is typically not sampled by many standard POC collection approaches, so that the net OC export of many headwater systems may be underestimated.

A key uncertainty in assessing the overall impact of steepland landsliding on headwater carbon balances is the fate of the mobilised carbon. Ramos Scharrón et al. (2012) recorded over 70% of landslide sediment from steeplands in Guatemala being delivered to higher-order streams, greater than the proportion in the two New Zealand studies discussed above. Clearly, the proportion of colluvial storage varies with landscape type and frequency of landsliding. Carbon storage potential of colluvial sediments, in-stream processing of particulate carbon, and the degree of enhanced carbon sequestration on disturbed sites, all impact the overall carbon dynamics. Ramos Scharron et al. modelled the global carbon balance for tropical steeplands varying between a source of 0.27 PgC yr^{-1} and a sink of 0.18 PgC yr^{-1}, based on extreme ranges of assumptions around the fate of carbon mobilised by mass wasting.

Further uncertainty around the fate of POC exported from steepland catchments relates to the initial source of the POC. Total POC fluxes from steepland headwaters may include both modern carbon derived from catchment soils and vegetation, but may also include sedimentary POC derived from sedimentary rocks in the catchments. Some studies have estimated that aged sedimentary POC can be the dominant POC component (Gomez et al. 2010). This is particularly likely to be the case where deep gully erosion of bedrock dominates over shallow landslide erosion of soils and biomass. Leithold et al. (2006) showed for a series of steepland catchments in Oregon, California and New Zealand that the ^{14}C age of POC correlated with sediment yield. Lower yields were dominated by young POC derived from shallow landsliding, whereas catchments where deep gullying produced higher sediment yield, had a higher proportion of old POC derived from bedrock.

Rates of POC export from forested headwaters are also substantially affected by changes in land use, in particular by exploitation of the forest resource. Deforestation typically accelerates POC loss, both because logging typically leaves large amounts of fresh but unrooted biomass at the surface, and also because deforestation is associated with slope instability and enhanced export of landslide sediment. In New Zealand, Gomez and Trustrum (2005) demonstrated that floodplain sedimentation associated with enhanced slope erosion due to deforestation in the 1920s was significantly enriched in organic carbon. However, they also noted that only 3–11% of the organic sediment flux was deposited on catchment floodplains, so that the majority of this enhanced flux is exported to the ocean. The fact that within-catchment storage may be limited in steepland systems has been widely noted (although rarely quantified). Headwaters are a source of POC and it has been argued (e.g., Scott et al. 2006) that because of both this and limited within-catchment storage, where river systems are short and the headwaters are well linked to the ocean, small mountain streams discharge disproportionate amounts of carbon to the oceans.

Scott et al. estimate that in New Zealand the average flux is 10 ± 3 tC km^{-2} a^{-1}. Figure 5.3 shows mapping of these estimates and it is apparent that the highest areal yields are associated with short, steep streams on the mountain front of Western New Zealand. Recently, Leithold et al. (2016) have argued that this is a general pattern with the highest POC fluxes associated with smaller steepland catchments on active plate margins, whilst lower fluxes result from more extensive within-catchment storage in large river systems situated on passive margins. The argument that headwater POC export is more effectively stored (or processed) in longer and larger river systems is supported by observations of Leithold et al. that POC age tends to be older in larger systems, indicating the processing of young labile carbon in the fluvial system.

Although there is good evidence that headwaters are important sources of organic matter to the wider river system, they are not passive conduits for the lateral transfer of eroded biomass and soil. One of the great uncertainties in

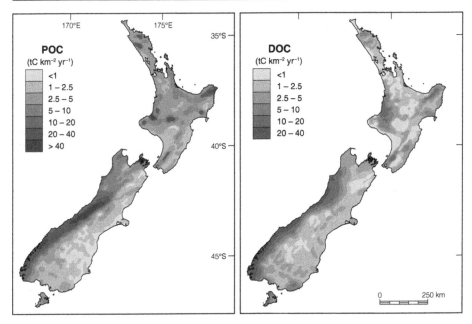

Figure 5.3   High POC yields in steeplands adjacent to the ocean. Source: After Scott et al. 2006. Reproduced with permission of Wiley.

assessing carbon flux from steepland headwaters is the fate of eroded carbon in colluvial storage. Data suggests that this may represent on the order of 50% of mobilised material (Hilton et al. 2011b; Page et al. 2004). Analysis of floodplain deposits downstream of eroding peatlands in the UK has demonstrated both the preservation of carbon at century timescales, but also indicates that floodplains are potential sites of carbon turnover (Alderson et al. 2019a). However, very little work on rates of decomposition or preservation of organic carbon in geographically and geomorphologically diverse floodplain sediments exists; this is a significant research need in this area.

One environment where significant carbon storage can take place within the fluvial system is in forested systems, where large organic debris plays an important role in both carbon storage and in the fluvial geomorphology of the system. Beckman and Wohl (2014) report significant carbon storage on floodplain systems, and within the channel, driven by the development of channel-spanning logjams in old growth and disturbed headwater forest systems. Carbon storage in logged systems, where there is a smaller amount of large organic debris, decreased by an order of magnitude. In a study of floodplain carbon storage in forested headwaters in Colorado, Wohl et al. (2012) identify unconfined floodplain segments as preferential areas of carbon accumulation because of the potential for complex log jams to provide storage zones in these systems. Jams create upstream storage zones, but also increase floodplain connectivity by forcing flow out of the channel across the floodplain, enhancing the deposition of wood and POC. Furthermore, because a dominant runoff generating mechanism in these systems is snowmelt, they do not experience frequent flash floods, and so tend to retain fallen wood on floodplain surfaces. Here too, carbon storage was also found to be greater in more complex channel forms associated with undisturbed old growth forest. Total carbon storage in these segments was equivalent to approximately 1700 years of net primary productivity in these systems, and radiocarbon ages on stored wood demonstrate long-term preservation. In this study, the largest component of the floodplain carbon storage is large fallen wood. Accumulation of large amounts of wood in the unconfined valley segments, together with characteristically high water tables, and hence lower decomposition rates at these sites, mean that total carbon storage can be substantial. Wohl et al. suggested that the 1% of the catchment area represented by headwater floodplains stored 25% of catchment carbon.

Although individual headwater catchments tend to be small, in combination they represent a significant proportion of total catchment stream length, so taken together with storage and decomposition of carbon in colluvial deposits, it is likely that steepland forest headwaters are not simply sources of OC, but sites of active turnover and storage of soil and biomass-derived carbon.

Carbon Cycling in Glaciated Headwaters

High elevation headwater systems are influenced by cryogenic processes and commonly support alpine glaciation. Glacial systems are not obvious landscape types to consider in the context of carbon cycling, since low temperatures limit both primary production and decomposition of organic matter. Glaciers are major causes of physical erosion and are very effective producers of fresh reactive fine-grained mineral material (Anderson 2007). Despite this fact, rates of silicate weathering in glacial systems are below the global average due to low temperatures. The most significant component of chemical weathering in glacial environments is carbonate dissolution, even in non-carbonate lithologies (Anderson 2007). However, where glacial sediments are deposited downstream, the high surface area and reactivity of the glacial flour can lead to elevated rates of silicate weathering. Anderson argues that as sites become more distal from active glaciation, carbonate dissolution reduces and silicate weathering increases, perhaps under the influence of vegetation establishment, promoting weathering of the fine-grained silicate material.

At geological timescales, Foster and Vance (2006) argue that enhanced silicate weathering associated with glacial erosion is a potential positive feedback in the climate system drawing down CO_2 from the atmosphere and enhancing cooling. In contrast, Torres et al. (2017) identify enhanced rates of sulphide mineral weathering from a compilation of water quality data from glaciated terrain. They argue that under glacial conditions the enhancement of sulphide weathering is greater than enhancement of silicate weathering, since weathering rates are greater for sulphide minerals. Torres et al. therefore argue that glaciation has the potential to release CO_2 through the action of sulphuric acid on carbonate minerals, creating a potential negative feedback so that glacial conditions may be self limiting. Horan et al. (2017) similarly have shown that the oxidative weathering of rock-derived organic carbon is 2–3 times higher in glaciated valleys in the New Zealand Alps, providing further evidence that headwater glaciation may drive hotspots of long-term CO_2 release due to enhanced weathering.

More work is required to understand the impact of glaciation on the slow carbon cycle both at a global level and in terms of understanding the geography of glacial weathering, but it is clear that rapid weathering in glacial systems is an important driver of change in the geological carbon cycle.

Carbon cycling in glacial systems is not however limited to long-term cycling of inorganic carbon. The ready supply of fine mineral material and the high rates of deposition associated with distal glacial deposits can also lead to preservation (by mineral complexation) and burial of locally derived organic matter (Anderson 2007), so that glaciation drives downstream carbon sequestration.

Recent work has shown that glacial systems are not biological deserts, but can maintain significant microbial populations. Barker et al. (2006) identified high degrees of variability in space and time of DOC quality in glacial environments in Antarctica, which they interpreted as indicative of microbial cycling of organic carbon in glacial systems. Similarly, Singer et al. (2012) showed in a study of 26 glaciers in the European Alps, that DOC quality was highly variable, and that a significant proportion of the measured compounds were labile, so that bioavailable DOC from glacial systems influences downstream carbon cycling in the fluvial system.

A variety of ecological niches exist on the glacier surface, including supraglacial sediment accumulations such as medial moraines of kame deposits and cryoconite holes in the glacier surface (Stibal et al. 2008). In supraglacial sediments, Stibal et al. (2008) found relatively low microbial abundance with bacteria, cyanobacteria and algae favoured by fine sediments with high water content.

Cryoconite is a dark coloured aggregate of microbial and mineral material found on the glacier surface. It is typically between 2 and 18% organic (Cook et al. 2016). Cryoconite holes are water filled depressions on the glacier surface, often cylindrical in form. These are believed to form because the darker colour of the cryoconite enhances melt due to the lower albedo of the material, and possibly also because of small additions of metabolic heat from microbial communities in the cryoconite mass (MacDonell and Fitzsimons 2008). Cryoconite has been widely studied and recent reviews cover the hydrology, formation, biology and biota of these features (e.g., Cook et al. 2016; Kaczmarek et al. 2016; MacDonell and Fitzsimons 2008). Cryoconite can be widespread on glacial surfaces covering between 0.1 and 10% of a glacier surface, with individual holes usually less than 1 m across and up to 0.5 m deep (Anesio et al. 2009).

The dark colour of cryoconite is due to the presence of humic substances which are the products of bacterial composition of algal material (Takeuchi et al. 2001). This implies that carbon is both being fixed and cycled within the cryoconite on glacier surfaces. Anesio et al. (2010) however argue that only 7% of the annual carbon accumulation in cryoconite holes is decomposed by heterotrophic bacteria, so that the cryoconite becomes a site of carbon accumulation on the glacier surface.

Whether cryoconite accumulations are net sources or sinks of carbon is still an active area of research. Telling et al. (2012) demonstrate that over half the variance in NEE in cryoconite that they studied in Arctic systems could be explained by the thickness of the cryoconite layer. Thicker sediment accumulations were autotrophic. However, spatial patterns of heterotrophy and autotrophy in cryoconite have also been explained with reference to nutrient availability, solar radiation levels and the nature of the glacier surface hydrology (Cook et al. 2016). Telling et al. (2012) showed that for the majority of sites they studied, rates of autotrophic carbon accumulation were not sufficient to account for observed cryoconite carbon stocks. Metagenomic analyses of cryoconite microbial diversity from an alpine glacier in the Austrian alps (Edwards et al. 2013) suggest a dominance of heterotrophic activity, with allochthonous sources of organic carbon such as moss fragments blown from ice marginal locations being particularly important sources. Allochthonous carbon sources are therefore potentially important in supporting microbial ecosystems on the ice. Issues around carbon source are explored in detail in Cook et al. (2016) and Stibal et al. (2012). Under some circumstances, heterotrophic bacteria may be supported by photosynthetic activity within the cryoconite hole, but delivery of organic matter blown by the wind or ancient OM melted from the glacier may also be significant sources of carbon for metabolism.

Anesio et al. (2009) argue that glaciers are largely autotrophic systems and estimated global fixation of carbon by cryoconite on glacial surfaces at 64 000 t a^{-1} (representing a balance of 98 000 t of primary productivity and 34 000 t of respiration). These are not large carbon fluxes. On an areal basis, Hodson et al. (2007) estimated CO_2 fluxes of just 12–14 kgC km^{-2} a^{-1} from cryoconite on glaciers in Svalbard. Stibal et al. (2008) note that, in relation to a glacial system on Svalbard, microbial primary productivity may be negligible relative to the flux of allochthonous organic matter, so that the role of the glacial system in the wider carbon cycle is as a store and potentially a source of recycled allochthonous organic material. Nevertheless, the extensive research on supraglacial biology does clearly show that these systems support a diverse microbial system, which is actively cycling carbon.

Glacial biology and carbon cycling is not confined to the glacier surface. Microbial life has been detected at low levels in deep glacier ice. Microbes sourced from atmospheric deposition or from surface biological activity may become incorporated into deep ice (Xiang et al. 2009), yielding distinctive microbial communities (Simon et al. 2009).

Subglacial sediments are also sites of biological activity. Skidmore et al. (2000) established that debris rich ice at the base of an Arctic glacier contained culturable microbes, which demonstrated metabolic activity at temperatures above freezing, consistent with warm-based glacial systems. Active microbes included methanogens, as well as nitrate and sulfate reducing organisms and aerobic chemoheterotrophs. Skidmore suggested that the carbon source for metabolic activity at the glacier bed might be pre-glacial organic sediments overridden by the glacial ice. Sharp et al. (1999) similarly argued that subglacial bacterial populations in basal ice are an order of magnitude greater than ice nearer the surface, and these populations have been shown to be active in processing carbon beneath the ice sheet (Foght et al. 2004; Sharp et al. 1999). Sharp et al. also argue that microbial processes may drive redox weathering reactions at the ice-sediment interface. This work was further developed by Skidmore et al. (2005) to assess microbial communities from ice overlying different lithologies, demonstrating that sub-glacial microbial communities vary with subglacial water chemistry, which suggests that microbial activity might contribute to subglacial weathering and consequent subglacial solute fluxes.

Measurement of fluvial carbon flux below glacier snouts integrates the supraglacial and subglacial biological processes described above, to give an integrated assessment of the role of glaciers and ice caps in carbon cycling. Hood et al. (2015) adopt this approach in a compilation of available data on DOC flux from glacial systems, and compare these data to estimates of total glacial storage of organic carbon. They estimate that globally, glacial runoff releases 1.04 (±0.18) million tonnes of dissolved carbon to the fluvial system, with 56% of this carbon derived from mountain glaciers. In contrast, estimated global englacial storage is 4.48 (±2.41) gigatonnes of carbon, of which 93% is stored in the Antarctic ice sheet (although it should be noted that this includes only englacial storage). The discrepancy between carbon storage and flux is explained by the relative rates of mass turnover in mountain glaciers and the Antarctic ice sheet. Hood et al. estimate that 80–90% of the DOC flux results from mass turnover, with the remainder representing a release of carbon from long-term storage due to net glacial wasting. This study also assessed POC storage and flux from glaciers; here the data are sparse and more variable, but POC storage is estimated to be 1.39 gigatonnes of carbon with an annual flux of 1.97 million tonnes. For POC, the flux figures are dominated by release from the Greenland ice sheet. A key observation by Hood et al. is that whilst glacial carbon fluxes are relatively small, laboratory incubations suggest that the carbon released is highly labile, with 25–95% of carbon being bioavailable. Aging of glacially derived OM has shown that it demonstrated much older ^{14}C ages than typical terrigenous material (Hood et al. 2009). This study also concludes that the material was dominantly microbially-sourced, consistent with turnover of OM stored in the glacial system by hetero-trophic microbes. Glacial DOM is therefore a distinctive input to the fluvial sys-tem, and carbon release from glacial systems is likely to impact downstream food webs to be rapidly mineralised and released to the atmosphere.

Linkages between the glacial system and the immediate pro-glacial environ-ment are important in controlling carbon cycling at a range of scales. The impor-tance of distal deposits as carbon stores and of glacial DOC as a source of carbon to downstream ecosystems have been identified above.

More generally, in headwater systems, disturbance and recovery are key con-trols on carbon dynamics, and in the case of the glacial system, the primary disturbance regime comes from the advance and retreat of the ice front. Where glacier advance overrides vegetated glacial forefields, organic matter is incorpo-rated into the basal ice and sediment layers of the glacier. Evidence of microbial activity at the base of warm-based glaciers means that there is potential that this old soil organic matter is a source of mineralised carbon during glacial periods (Sharp et al. 1999; Skidmore et al. 2000) and old glacial carbon has been demon-strated to be a component of downstream food webs in glacial systems (Fellman et al. 2015).

Glacial retreat exposes fresh sediments and bedrock from the glacier bed. These systems, initially, are largely abiotic, but the primary succession of these sites and subsequent carbon accumulation has been well studied because of the well-constrained chronosequences, which can be defined on glacier forefields during glacier retreat. Such studies typically show rapid development of vegeta-tion cover and increasing carbon storage during the early stages of soil formation (Guelland et al. 2013; Kabala and Zapart 2012; Thomazini et al. 2014) and the succession of microbial communities (Bacteria, Archaea and Fungi) has been demonstrated to correlate with changing forefield carbon and nutrient contents (Zumsteg et al. 2012).

In an excellent example of the potential of this approach, Guelland et al. (2013) characterised a chronosequence spanning 7–128 years of post glacier retreat at a site in the Swiss Alps. Soil CO_2 flux increased from 9 to 160 gC m^{-2} a^{-1} as time since deglaciation increased, and soil depth and vegetation cover increased. Associated with changes in carbon flux, were changes in the nature of the carbon source. At the youngest sites, ^{14}C dating suggested that ancient (pre-glacial) organic matter was the dominant carbon source. At sites of intermediate age (circa 50 years), modern organic carbon sourced from forefield vegetation dominates the gaseous carbon flux. The oldest sites were accessing slightly aged carbon, which Gelland et al. attributed to heterotrophic respiration of older SOM formed since deglaciation.

A potential link exists between the initiation of succession and carbon fixa-tion on glacial forefields and the biology of the retreating glacier. Some authors have argued that organic matter and microbial life from englacial and subglacial sources are an important factor in initiating biological succession on recently exposed glacial surfaces (Kaštovská et al. 2005, 2007). It has been noted that the development of active nutrient cycling on successional sites can precede the development of plant cover and that microbial sources dominate the organic matter content of early successional stages (Schmidt et al. 2008). Bardgett et al. (2007) demonstrated at a glacier forefield site in the Austrian Alps that heterotrophic microbial communities were utilising old carbon at the youngest deglaciated sites. It is possible, therefore, that in newly deglaciated terrain, both the microbial community and the carbon substrate that it exploits, reflect an inheritance from the glacial system and potentially from pre-glacial ecosystems.

Carbon cycling in glaciated headwaters has been intensively studied, particularly in the last five years. Cryospheric environments would normally be associated with low rates of metabolic activity and limited carbon turnover. The extensive work on glacial microbial communities has demonstrated that this is not the case. Pre-glacial organic matter in basal debris, cryoconite environments on the glacier surface and even englacial environments have been demonstrated to support both autotropic and heterotrophic microbial life. In common with other headwater systems, the downstream export of organic matter from glacial systems is significant in supporting downstream ecosystems, and the dynamism of the glacier fronts leads to significant disturbance and recovery of glacier forefield environments.

Overall, the magnitude of storage and export of organic matter from glaciers is small in the context of wider global biomes, and it is perhaps ironic that these systems are so well studied and understood. However, the observation that OM export from glacial headwaters is strongly bioavailable, and the fact that deglacial environments are increasing rapidly due to climatic change and glacier melt, mean that understanding of these systems is potentially increasingly important under the conditions of climate change. Hood et al. (2015) estimate a five-fold increase in release of DOC from glaciers to fluvial and marine systems by 2050, due to loss of glacier mass; 63% of this additional flux is estimated to come from headwater mountain glacier systems. The DOC flux is likely to be mineralised and constitute a release of carbon to the atmosphere. The fate of comparable changes in the POC flux will be a function of the nature of the downstream storage and processing of this sediment. Glacial forefields will be an important locus for storage and cycling of this material.

Carbon Cycling in Headwater Peatlands

Global peatland cover is estimated at circa 4.2 million km² of which the northern temperate peatlands comprise around 80%, with the remainder being tropical peatlands (Limpens et al. 2008; Xu et al. 2018). Peat accumulates at sites where the water table is close to the surface most of the time. One major classification of peatlands is by the source of the water contributing to the water table (Charman 2002; Evans and Warburton 2007). Ombrotrophic peatlands (bogs) are those fed entirely by rainwater with no significant inflows, in contrast to minerotrophic systems (fens) where high water tables are maintained by inflowing water. Given that this is a topographically-based (geomorphological) classification, a reasonable starting point for an estimate of headwater peatland cover would be the area of ombrotrophic peatlands. Of the approximately 4.2×10^6 km^2 of temperate peatland cover, 1×10^5 km^2 is classified as upland ombrotrophic peat (Lindsay et al. 1988). This is likely to be a minimum estimate for all headwater peatlands, since peatlands form in a range of upland contexts (Figure 5.4). Minerotrophic conditions can occur in headwater systems in areas such as valley mires formed in cirque basins, and on low-angled alp benches in the high mountains (Figure 5.5a). Similarly, headwater systems exist in areas of low relief and in the mixed fen and bog landscapes of the Canadian and Siberian shields. Extensive headwater peatlands occur in blanket peatlands and sloping mire systems, where peat forms in areas of high precipitation on relatively steep slopes (e.g., Figure 5.5b).

Overall, peatland systems are major sites of accumulation of terrestrial carbon. Estimated total carbon storage in northern peatlands is on the order of 450 Pg (Gorham 1991). On this basis, proportional carbon storage in upland headwater systems would be circa 15 PgC, which as noted would be a highly conservative estimate.

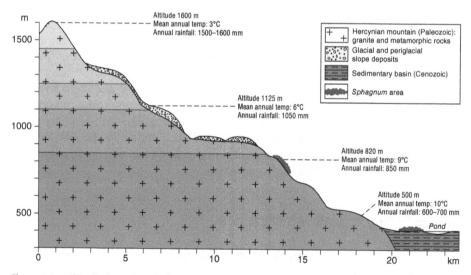

Figure 5.4 Altitudinal zonation of upland mire types in the uplands of Western Europe: the example case being the Monts du Forez mountain range in France. Source: After Cubizolle and Thebaud-Sorger 2014 (http://mires-and-peat.net/pages/volumes/map15/map1502.php) Licensed by CCA 4.0.

Figure 5.5   Headwater peatlands. (a) Blanket peatland in the North Pennines, UK. Source: Martin Evans. (b) High elevation Andes peatland, Ecuador. Source: After Chimner and Karberg 2008. Courtesy of Rodney Chimner.

Headwater peatlands have been well studied with respect to their hydrology and geomorphology (e.g., Cowley et al. 2016; Evans and Warburton 2007; Fryirs et al. 2016; Holden 2005; Holden and Burt 2003), in part because, unlike higher-order peatland catchments, they are more likely to have 100% peat cover and so provide a suitable system for the study of peatland processes.

Geomorphological study of peatlands can be classified into two areas of focus. Geomorphological analysis of intact peatlands with complete vegetation cover and limited sediment mobility has focussed on the relation between landscape topography and peatland type (e.g., Charman 2002). Whereas in degraded peatlands, where degradation of vegetation cover leads to active erosion of the peat mass, research into the processes and forms associated with peat erosion has dominated (e.g., Evans and Warburton 2007; Li et al. 2018).

Because peat soils are typically 90% organic, the accumulation and erosion of the peat mass defines the geomorphology of the peatland landscape, as well as playing the dominant role in the peatland carbon budget. In peatland systems, carbon cycle dynamics, the role of microbial processes in processing terrestrial carbon and the geomorphology of the peatland, are strongly interdependent.

Geomorphology and Carbon Cycling in Intact Peatlands

Peatlands across the globe exhibit considerable variation in ecology and form and have been classified in a wide variety of ways (Charman 2002). Maintenance of high water tables is fundamental to peatland function and so hydrology and morphology underpin most peatland classifications (e.g., Figure 5.6).

Figure 5.7 represents the carbon balance of a peatland system. Accumulation of carbon and vertical growth of the peatland are supported by high rates of

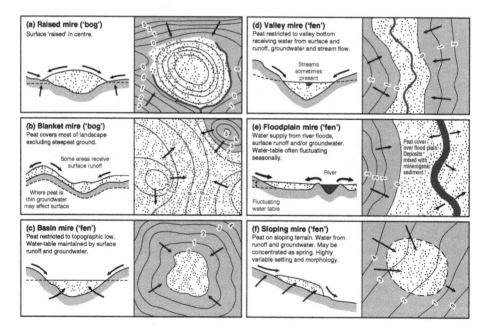

Figure 5.6   Hydro-topographical classification of mire types. Source: After Charman 2002. Reproduced with permission of Wiley.

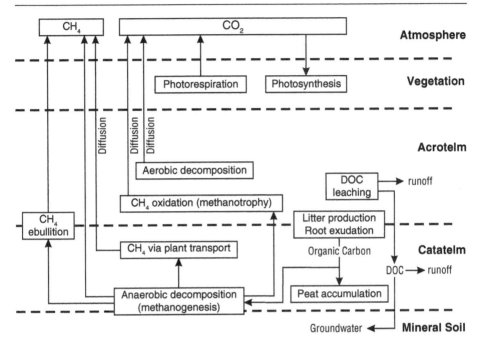

Figure 5.7   The Peatland carbon cycle. Source: After Joosten et al. 2016. Reproduced with permission of Cambridge University Press.

biomass accumulation and by the suppression of litter decomposition. Rates of microbial decomposition in ombrotrophic peatlands are limited by high water tables (creating anaerobic conditions), by acidity and by nutrient poor conditions.

Peatland form is significantly impacted both by topographic controls on peatland water balance, but also by rates of peat accumulation. Peat depths are commonly 2–3 metres in areas of blanket peat and can reach tens of metres in topographic depressions or in raised bogs. At the local scale, peatland form can be controlled by a complex interaction of primary productivity, peat hydrology and organic matter decomposition. In some conditions (e.g., raised bogs) these processes drive a characteristic bog morphology (Ingram 1982). In extensive peatlands, different peatland types coalesce so that peatland form is both, controlled by, and is a control upon, landscape morphology. These complex mosaics of habitat type support a diversity of plant-microbe assemblage ages, which are fundamental to controlling rates of carbon accumulation in the landscape (Andersen et al. 2013).

Low rates of organic matter decomposition in peatlands have commonly been ascribed to physical conditions, particularly as noted above, the cool, acidic and anaerobic conditions which prevail below the peat surface. Whilst strong empirical relations can be found between; for example, water table and CO_2 or CH_4 flux

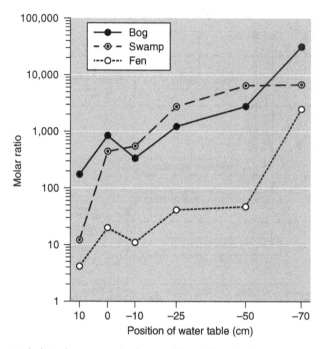

Figure 5.8   Empirical relation between peatland water tables and flux of methane and carbon dioxide from peatland surfaces to the atmosphere. Source: After Moore and Knowles 1989. © Canadian Science Publishing or its licensors.

(derived from OM decomposition) at the peatland surface (Figure 5.8), the nature of the relations varies significantly between peatland types (e.g., Sulman et al. 2010), and the process basis of these relations has been less well understood. Freeman et al. (2004) demonstrated the importance of a specific extracellular enzyme (phenol oxidase) in controlling litter decomposition in peatlands. Phenolic compounds accumulate in decomposing organic matter and inhibit the operation of hydrolase enzymes, which play a key role in OM decomposition. Decomposition of phenolic compounds by phenol oxidase reduces this inhibition and so promotes the breakdown of cellulose. Phenol oxidase activity is dependent on the presence of oxygen, and so in saturated anaerobic conditions in deep peat, accumulation of phenolic compounds significantly reduces decomposition rates, promoting peat growth.

Production of extracellular enzymes which drive the decomposition of organic matter in peatlands, and which therefore control peatland carbon cycling, is a function of soil biota so that in peatlands, as in all soil systems, a process-based understanding of carbon cycling must be underpinned by analysis of microbial dynamics.

Microbial Processes in Intact Peatlands

Peatlands accumulate organic matter and are significant carbon stores because litter is decomposed at a rate less than the rate of net primary production. Therefore, although rates of primary productivity in cooler upland conditions may be below those of lower elevations, net carbon sequestration is still significant. Typical rates of carbon storage in upland blanket peatlands in pristine conditions are circa 55 tC km^{-2} a^{-1} (Worrall et al. 2003). Intact upland peatlands therefore fix carbon and accumulate thick organic peat layers which are a substantial component of terrestrial carbon storage. Decomposition of organic matter in peatlands occurs through microbial breakdown of plant material to simpler organic molecules. These simpler molecules fuel microbial respiration and consequent release of gaseous carbon to the atmosphere (Figure 5.8). Consequently, a mechanistic understanding of peatland carbon cycling is closely tied to understanding microbial community dynamics and function.

Fungi decompose organic matter through the production of extracellular enzymes and are commonly regarded as the primary decomposers of plant matter in peatlands, whereas bacteria often occupy niches adjacent to fungal hyphae in order to access smaller organic molecules, which are products of fungal decomposition (Gilbert and Mitchell 2006). Studies of peatland OM quality by Moody et al. (2018) suggest that fungal degradation of lignin is a key process in the production of DOM in peatlands.

Studies of peatland microbial diversity show considerable variation. Golovchenko et al. (2007) studied peatlands in Tver Oblast, Russia and showed that in nutrient poor ombrotrophic peatlands fungal biomass exceeded bacterial biomass (55–99% of biomass), whereas in minerotrophic systems bacterial biomass was 55–86% of the total. They noted however that bacterial activity exceeded fungal activity. Similarly, Winsborough and Basiliko (2010) analysed microbial activity (respiration rates) for a range of peatland environments and showed that bacterial processes are dominant with fungal–bacterial activity ratios ranging between 0.31 and 0.68.

In the Massif Central region of France an analysis of microbial food chains in peatlands has shown microbes are supported by decomposition of organic matter. Heterotrophic bacteria (15% of microbial biomass) feed on dissolved organic matter (DOM) and are in turn metabolised by heterotrophic protists, such as testate amoebae (48% of microbial biomass) (Gilbert et al. 1998). The study concluded that heterotrophic microorganisms were relatively more important in peatlands than in other aquatic environments. Gilbert et al. did not explicitly consider fungi in their study, although fungal decomposition of litter is the likely source of DOM at the base of the food chain. Similar results have been reported by Mitchell et al. (2003) in a study of five sphagnum peatlands across Europe, who found that heterotrophic organisms represented 78–97% of biomass, and again testate amoebae were a dominant protozoan species.

Rates of carbon turnover in peatlands are significantly controlled by water table and nutrient status, and the mechanism driving these responses is assumed to be changes in microbial community and activity in response to physical and biochemical change in the system. As noted above, the nutrient status of peatlands influences microbial community structure and only highly adapted species flourish in the harsh peatland environment. Oligotrophic peatlands consistently have lower bacterial diversity than more nutrient-rich systems (Andersen et al. 2013).

In raised mire systems, Hall and Hopkins (2015) demonstrated that ombrotrophic and minerotrophic parts of the mire system sustained similar microbial biomass, but both the rate of decomposition of litter and the rate of microbial respiration was elevated in the more nutrient-rich minerotrophic parts of the mire. This is consistent with higher rates of respiration from bacteria-favoured, more nutrient-rich minerotrophic systems.

Peatlands are characterised by aerobic conditions in the acrotelm layer above the mean water table, and largely anaerobic conditions in the oxygen-depleted environment below the mean water table. Above the water table, microbial respiration of organic molecules releases gaseous carbon in the form of CO_2. Below the water table in the catotelm, highly adapted microbial communities respire anaerobically, with CH_4 as the gaseous by-product. These changes in microbial diversity and activity under water table control, underpin the well-established relation between mean water tables in peatland systems and the balance of gaseous carbon release as CH_4 or CO_2 (e.g., Moore and Knowles 1989 and Figure 5.8).

The biota of the catotelm is dominated by anaerobic bacteria and Archaea. Methanogenesis occurs in three steps controlled by species of bacteria and Archaea. These steps have been characterised by Kamal and Varma (2008) as: 1) bacterial hydrolysis in anaerobic conditions; 2) fermentation (bacteria and Archaea); and 3) methanogenesis (Archaea). In addition to methanogenesis in the catotelm, peatlands also support communities of methanotrophic bacteria. In a study of ombrotrophic peatlands in northern England, Edwards et al. (1998) identified 50 species of methanogens and methanotrophs in 19 genera. Methanogenic species were confined to the anaerobic zone, whereas methane oxidising species were found throughout the peat column. Methane oxidising potential was highest in the aerobic upper 10 cm of the peat profile. Andersen et al. (2013) suggest that water table lowering leads to an increase in fungal biomass, and an associated increase in gram negative bacteria, which feed on simple substrates. Shifts in water table also affect the balance of methanogenic and methanotrophic bacteria. Lower CH_4 emissions at low water tables are therefore a function of both lower anaerobic methane production and higher aerobic methane oxidation (both microbial) (Sundh et al. 1994). Associated increases in CO_2 flux are driven by increases in fungal decomposition and in methane oxidation.

Edwards et al. (1998) note the complexity of the communities they studied, with a variety of species dominant in different parts of a collected sediment core (i.e. different vertical positions relative to mean water table). Many of the species they identified were novel, a pattern also noted by Rooney-Varga et al. (2007) who found in a study of Alaskan peatlands that uncultivated methanogenic Archaea species dominated the biota, and that species mix varied strongly with changes in vegetation and environment.

Research in peatland microbiology has been thoroughly reviewed by Andersen et al. (2013), who conclude that peatlands have repeating mosaics of highly adapted organisms. A key observation is that peatlands are characterised by species sorting (not all microbial species are active at any time due to changing conditions) and functional redundancy (different organisms carry out the same function). Microbial communities respond to changes in water table, but also to changes in substrate, with evidence of succession in fungal communities in response to changes in litter quality. Species sorting and functional redundancy contribute to the resilience of peatland function; for example, in response to short-term water table drawdown, but longer-term changes in water table and vegetation can drive functional change (Andersen et al. 2013). Rates of microbial change in response to significant change in the peatland environment may be relatively slow. An important study of restored UK peatlands showed that degraded peatlands had low microbial diversity, and that whilst the short-term response of these systems to restoration was rapid, the microbial system continued to change over 25 year timescales (Elliot et al. 2015).

Overall, the microbial system of peatlands has been reasonably well studied, certainly in comparison to the other systems considered in this chapter. A common finding is the tight linkage of vegetation, soil environment and microbial communities. For this reason, understanding of peatland carbon dynamics has typically developed from analysis of change in vegetation and the soil environment. It has however engaged less closely with microbial change as the key driver of peatland carbon cycling. Critically, the links between microbial community dynamics and peatland function have not been systematically explored. In particular, the short-term response of microbes to rapid changes in physical conditions (e.g., rainstorms or drought) and microbial activity have not been well studied. Since microbial action drives OM decomposition and gaseous carbon flux from these systems, the dynamics of peatland carbon cycling are closely linked to the response of microbial activity to changes in soil environment and to changes in microbial communities at a range of time and space scales.

Geomorphology and Carbon Cycling in Degraded Peatlands

A second major focus in peatland geomorphology has been work on eroding peatlands. In peatland systems, the high organic content of soils (typically 90%)

means that the study of peatland geomorphology and peatland carbon cycling in these systems is intimately connected. Key areas of investigation have included:

1) The characteristic topography produced by distinctive erosional processes operating at the peat surface (Bower 1961; Evans and Lindsay 2010a; Li et al. 2018)
2) The role of organic sediment flux from rapidly eroding peatlands in modifying peatland carbon balance (Billett et al. 2010; Pawson et al. 2012)
3) The impact of modified topography on eroding peatlands on the plant–water table microbial interactions which control carbon accumulation (Evans and Lindsay 2010b)

Peatlands across the globe have experienced significant degradation through the processes of drainage, overgrazing and pollution (Joosten 2016). Across large parts of the world, peatland drainage for agricultural improvement has been the major impact, which has lowered water tables and increased CO_2 flux to the atmosphere. In upland peatlands, agricultural uses have typically been less intense, although in the UK extensive shallow drainage of blanket peatlands has contributed to the severe degradation of upland peats (Holden et al. 2004). Upland peats, subject to more severe climate and often forming on steeper slopes, are particularly susceptible to physical instability. Extensive erosion of peatlands has been reported in Australia (Pemberton 2005), in the peatlands of northern China (Zhang et al. 2016) and across the UK and Ireland (Evans and Warburton 2007).

The physical stability of peatlands is strongly dependent on the maintenance of an intact vegetation layer and high water tables. Where removal or weakening of the vegetation layer shifts the balance between eroding (wind, rain and runoff) and resisting forces (physical protection of the surface by vegetation and binding of the upper soil layer by roots), exposure of bare peat can lead to a loss of physical stability and extensive gully erosion (Evans and Lindsay 2010a, 2010b) (Figure 5.9).

Gully erosion of peatlands dramatically modifies the carbon balance of the peatland. The direct impact of erosion is the flux of particulate carbon. In the eroded peatlands of the UK, rates of particulate carbon loss can reach 74 t C km^{-2} a^{-1} (Pawson et al. 2008). Eroded peat deposited on floodplains at the margins of peatlands can be rapidly oxidised (C. Evans et al. 2013). Physical instability is also associated with melting permafrost, and can promote elevated particulate flux and export of old carbon through the fluvial system (Guo et al. 2007).

Gully erosion also impacts peatland carbon balance indirectly in two ways: 1) erosion of gully systems across intact peatlands causes local drainage and reduction in water tables, which leads to deepening of the aerobic layer and increase in CO_2 flux to the atmosphere; and 2) intense gully erosion has a significant impact on land cover, with unvegetated gully floors making up a significant part of the land area. These unvegetated surfaces do not fix carbon by photosynthesis, reducing

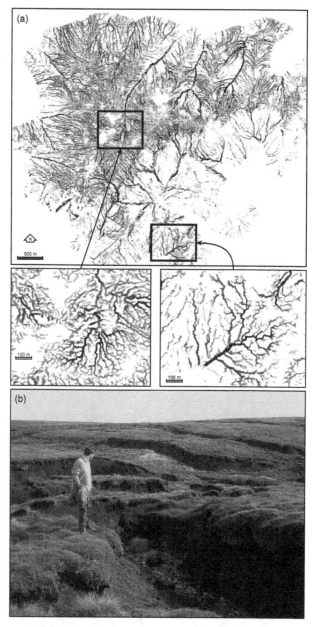

Figure 5.9   Gully erosion in the south Pennines UK. (a) Mapping of extensive gully erosion on the Bleaklow Plateau. Source: After Evans and Lindsay 2010a. Reproduced with permission of Wiley. (b) Severe gully erosion in the area mapped above. Source: Martin Evans.

uptake of atmospheric carbon at the ecosystem level. In the severely eroding peatlands of the south Pennines in the UK, the carbon loss from the peatland (Evans and Lindsay 2010b) showed that gully erosion had the potential to shift the peatland carbon balance from a net sink of carbon to a net source. The largest impact was removal of vegetation, with the irreversible impacts of gully drainage on water tables producing a lower proportion of total carbon loss.

Peatland degradation leads to the significant export of particulate and dissolved carbon from the peatland system (Figure 5.10), as well as enhancing carbon loss to the atmosphere. The net effect on terrestrial carbon cycling is in large part dependent on the off-site fate of these carbon sources. Peat re-deposited onto floodplains can be rapidly oxidised (C. Evans et al. 2013, see also Chapter 7 in this book), but some of this material is also stored for long periods (Alderson et al. 2019a). In the fluvial system, peat is also actively cycled in-stream, with high POC concentrations being linked to DOC production and CO_2 flux from the stream (Brown et al. 2019).

Peatland degradation also impacts on the microbial communities which drive carbon cycling. Studies of these impacts are limited, but in the English Peak District, it has been shown that bare peat areas have lower numbers of cultivable microbes and distinctive communities (Elliott et al. 2015). Changes in microbial populations in response to disturbance are commonly observed, but a proper understanding of the linked changes in microbial diversity and function is a future research need (Andersen et al. 2013).

The restoration of peatlands offers the potential to enhance the resilience of upland carbon stores however, whilst vegetation restoration and reductions in

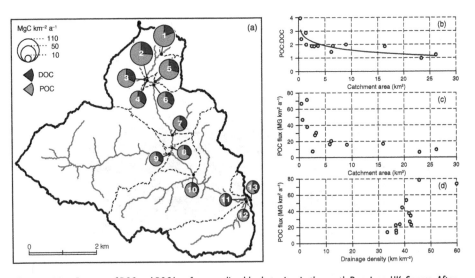

Figure 5.10 Patterns of DOC and POC loss from eroding blanket mires in the south Pennines, UK. Source: After Pawson et al. 2012. Reproduced with permission of Wiley.

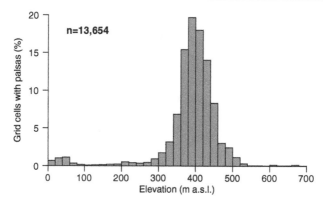

Figure 5.11   Altitudinal distribution of eroding high altitude Palsa mires in northern Norway. Rapid lateral erosion of the Palsa mires is assumed to be linked to thawing of permafrost due to climate change. Source: After Borge et al. 2017 https://tc.copernicus.org/articles/11/1/2017/tc-11-1-2017.pdf. Licensed by CCA 4.0.

particulate carbon flux (Shuttleworth et al. 2015) can be achieved in 5–10 year timescales, but full restoration of microbial function timescales may exceed 25 years (Elliot et al. 2015). Stable peatlands are a major store of carbon in upland systems, and so diverge from the notion of headwaters as net exporters of organic matter. However, geomorphological instability under conditions of human impact or climate change (e.g., Figure 5.11 and Li et al. 2017b) can lead to dramatic shifts in this balance, so that headwaters become a major source.

Understanding Carbon Cycling in Headwater Peatlands

Because of the carbon density of peatland soils, and their considerable depth and extent, small changes in peatland function can significantly impact their role in the terrestrial carbon cycle. In order to understand the resilience of these important carbon stores in response to future climate change, there is a need to more clearly understand the processes underpinning peatland carbon dynamics. Two key drivers of change in the peatland carbon cycle are changes in water table and erosion. Understanding characteristic timescales of change and recovery in these parameters, and the microbial responses to these, are critical. The processes and the microbial biota are reasonably well understood, but the study of their interaction is in its infancy and should be a key research agenda for this area of science.

Headwater Geomorphology and the Terrestrial Carbon Cycle

Table 5.2 summarises some typical rates of carbon flux reported in the literature for headwater systems. It is clear that both peatlands and steeplands are potentially significant sources of carbon to downstream systems. The magnitude of these fluxes is such that they have the potential to significantly modify

Table 5.2 Typical values of headwater carbon flux reported in the literature.

Headwater context	Carbon flux	Notes	Citation
Landslide transport in New Zealand forested steeplands	7.6 tC km^{-2} a^{-1}	Total estimated landslide transport	(Hilton et al. 2011b)
New Zealand grassland steepland headwater landsliding	50 tC km^{-2} a^{-1}		(Page et al. 2004)
Tropical steepland mass wasting	0–39 tC km^{-2} a^{-1}		
Tropical Montane steepland carbon balance	−25 − + 38 tC km^{-2} a^{-1}	Calculated from model scenario estimates in paper	(Ramos Scharrón et al. 2012)
Global glacial DOC flux	0.07 t km^{-2} a^{-1}	Based on data in Hood et al. and assumed global glacial area of 1.5×10^7 km^2	(Hood et al. 2015)
Global glacial POC flux	0.13 t km^{-2} a^{-1}	As above	(Hood et al. 2015)
POC flux from eroded peatlands	Up to 78 t km^{-2} a^{-1}	Data from severely eroded peatlands in S. Pennines	(Pawson et al. 2012)
DOC flux from eroding peatlands	Up to 28 t km^{-2} a^{-1} (temperate) up to 93 t km^{-2} a^{-1} (tropical)	Data from S. Pennines and from SE Asian peat swamp forests	(Moore et al. 2013; Pawson et al. 2012)
Typical intact upland peatland carbon balance	Sink of 55 t km^{-2} a^{-1}		(Worrall et al. 2003)

local and regional carbon balances. For reference, the IPCC estimate of 0.9 PgC flux of fluvial carbon from the land to the ocean is equivalent to 6 t km^{-2} a^{-1} when averaged across the global land area (Cubasch et al. 2013). Glacial systems are a small component of the global system, although as noted above they may be locally important carbon sources in some headwater systems, and glacial forefield systems are likely to become more significant under conditions of climate change.

Figure 5.12 presents a headwater carbon landsystem model, which summarises the key geomorphological contexts for headwater carbon cycling discussed in this chapter. Headwaters are high energy environments, with active reworking of soils and sediments by geomorphological processes. Headwaters account for the majority of the stream length in the fluvial system, and this together with the active nature of the environment, means that they are often characterised by efficient transfer of material from hillslopes to channels.

Delivery of Organic Matter to the Fluvial System

Conventional ecological theory holds that headwaters are sources of coarse particulate organic matter to stream systems. High rates of POC flux identified across all three environments support this conceptualisation, although the nature

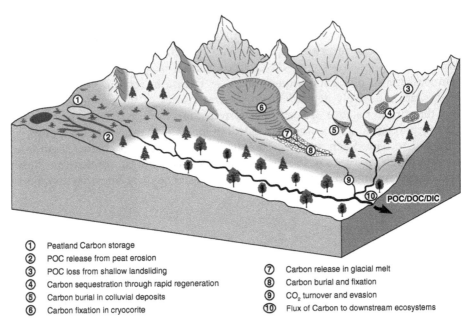

① Peatland Carbon storage
② POC release from peat erosion
③ POC loss from shallow landsliding
④ Carbon sequestration through rapid regeneration
⑤ Carbon burial in colluvial deposits
⑥ Carbon fixation in cryocorite
⑦ Carbon release in glacial melt
⑧ Carbon burial and fixation
⑨ CO$_2$ turnover and evasion
⑩ Flux of Carbon to downstream ecosystems

Figure 5.12   Characteristic carbon landsystems of headwaters. Source: Martin Evans.

of the OM (tree trunks vs. peat particles) can vary significantly between environments. However, the standard understanding treats the terrestrial catchment as a black box. Understanding the geomorphological controls on the delivery of OM to the fluvial system, provides a framework for consideration of temporal patterns of OM delivery. Typically, this is episodic. In peatland systems, POM flux from eroding peat surfaces is driven at a synoptic scale by rainfall events. In glacial systems, OM is released by periods of melt so that seasonal and decadal patterns of melt are a key control. In steeplands, connectivity is maximised by landsliding during high magnitude storm events, with return periods likely to be years to decades.

OM delivered to the fluvial system is either stored, biologically processed within-stream, or transmitted as a source of OM to downstream systems. Understanding the temporal phasing of delivery provides a key context for the analysis of the fate of fluvial OM. OM which moves between stable stores can be regarded as climate neutral, whereas active cycling of carbon by heterotrophic fungi and bacteria, either in the aquatic system or in terrestrial sites of sediment deposition, leads to release of carbon to the atmosphere.

The Impact of Geomorphological Change on Carbon Cycling

Landslide disturbance, glacier melt and the onset of peat erosion are all examples of episodic geomorphological change in headwater systems, which not only directly control carbon flux, but also significantly modify local ecosystems and their potential for carbon sequestration. Timescales for decomposition of OM in storage, change in primary productivity post-disturbance and the recovery of microbial populations post-disturbance, can all be years to decades, so that temporal trends of carbon sequestration, or release in response to disturbance events, are sensitively dependent on the trajectories of these changes. Modelling the carbon balance of headwater systems, therefore, requires integration of geomorphological and ecological modelling in a way that has not been properly explored to date. Such an integrated approach will rely on characterisation of the headwater landsystem, stronger empirical data on soil–atmosphere gas fluxes and characterisation of the extent and the persistence of carbon-rich deposits in headwater systems.

Conclusions

The dynamic nature of headwater systems means that these are sites of carbon turnover. High rates of geomorphological disturbance to equilibrium ecosystems mean that rates of primary productivity and OM decomposition are tightly linked to geomorphological processes. More integrated work on the nature of

Table 5.3   Characteristics of headwater carbon cycling and geomorphological controls and an assessment of key research needs in this area.

Characteristics of headwater geomorphology and carbon cycling

- The importance of geomorphological processes as vectors for carbon transport and high degrees of connectivity between steep hillslope systems and dense drainage networks.
- Episodic delivery of POM to fluvial systems under geomorphological control
- Importance of heterotrophic bacteria and fungi in degradation of POM
- The role of geomorphological processes in disturbing ecological systems leading to
  - Successional changes in primary productivity
  - Changes in microbial diversity and gradual recovery
- Headwaters are significant sources of OM to downstream ecosystems
- Important role of carbon sequestration in upland depositional landforms (landslide deposits, sandur plains, headwater floodplains, peatlands)

Key research needs

- Direct measurements of gaseous carbon flux to the atmosphere from headwater systems (aquatic and terrestrial)
- Data from a wider range of environments on rates of recovery for microbial and macroscopic communities from disturbance
- Utilising data on recovery rates and disturbance regimes to model trajectories of change in terrestrial carbon sequestration potential for headwater systems

these linkages is urgently required to open up the black box of terrestrial carbon cycling in the headwaters. Key characteristics of headwater systems are identified in Table 5.3, together with a summary of the research agenda identified above.

Headwater systems are often steep, remote, and areas of severe climate. As such, they are challenging places to work and are often under-represented in monitoring networks. However, as geomorphologically active sites, they play an important role in carbon turnover and in the supply of organic matter to the source waters of the fluvial system. Headwater peatlands are a special case. Higher rainfall and cooler temperatures sustain peat growth in areas of low slope in headwaters and these locations are important carbon stores in the terrestrial system. Stable peatlands store large amounts of carbon at timescales in excess of millennia. However, where there is potential for these stores to become physically unstable, they have the capacity to become major sources of carbon to the fluvial system and the atmosphere. The dynamic and connected nature of headwater carbon stores means that a focus on understanding the physical and microbial processes driving change in these systems is an important area of study.

Chapter Six
Hillslope Soil Erosion and Terrestrial Carbon Cycling

Introduction

Cole et al. (2007) wrote of the importance of 'plumbing the global carbon cycle', emphasising the active role of freshwaters in the cycling of carbon. Drainage networks are an effective unit of analysis for understanding terrestrial carbon fluxes because the plumbing effectively integrates ecosystem processes across the catchment. However, the units that are 'plumbed' together, hillslopes, are not a static component of the system. Particulate carbon may be transported across hillslopes by erosion, and the erosion status of hillslopes can also influence vegetation patterns, and so, rates of primary productivity.

Hillslope erosion was one of the first elements of the terrestrial carbon cycle to receive significant attention in the geomorphological literature. In particular, in the field of agricultural soil erosion, traditional concerns about sustainable agriculture that considered soil organic matter content and soil erosion, meant that there was a community of soil scientists and geomorphologists equipped to address emerging concerns about terrestrial carbon storage.

Two excellent recent reviews cover much of this literature (Doetterl et al. 2016; Kirkels et al. 2014). This chapter will cover some of this material concisely, to provide context for a discussion aiming to integrate geomorphological and ecological understanding, and to develop a conceptual model of the hillslope carbon landsystem.

Natural soil systems are often considered to be in equilibrium, such that surface soil erosion is balanced by production of new soil material by chemical weathering at the base of the profile, and through addition of litter at the top

Geomorphology and the Carbon Cycle, First Edition. Martin Evans.

of the sequence. In contrast, agricultural systems subject to accelerated erosion of topsoil may experience net losses of soil and organic matter from hillslopes (Doetterl et al. 2016).

Globally, the World Bank reports just under 48.9 million km^2 of land or 37.5% of land area as in agricultural use (World Bank 2017), of which 11% is arable land. Unsurprisingly, therefore, much of the geomorphological work on hillslope erosion in the context of carbon cycling relates to agricultural systems. The exception is the work on tropical landslides, which has been covered in Chapter 5. The impact of agriculture on soil erosion rates is dramatic. Montgomery (2007) showed that modern plough-based agriculture can lead to increases in rates of hillslope soil erosion by 1–2 orders of magnitude over naturally vegetated systems.

Local impacts of agricultural activity can be identified in prehistory, for example, accelerated erosion due to Mayan agriculture as early as 700 BC (Anselmetti et al. 2007). Across Europe, substantial soil erosion associated with first Neolithic, and then medieval agriculture, can be identified (e.g., Kuhn et al. 2009). The most dramatic increases in agricultural erosion however are associated with expansion of intensive agriculture and the spread of mechanised agriculture from the twentieth century (e.g., Figure 6.1). These impacts can be dramatic, for example Trimble (1983) describes rates of floodplain sedimentation of up to 1.5 metres in eight years associated with intensification of upland agriculture in Wisconsin in the period up to 1930. The most significant impacts of hillslope erosion on terrestrial carbon cycling are therefore a relatively recent phenomenon, which may be considered part of the 'great acceleration' of anthropogenic impacts on the global system which occurred during the twentieth century (Steffen et al. 2015).

Hillslopes in the Terrestrial Carbon Cycle

Estimates of terrestrial organic carbon flux to the oceans vary widely. Data collated by Schlünz and Schneider (2000) show a range of estimates from 0.03–1 PgC yr^{-1} (mean 0.4; standard deviation 0.23 n = 19) and they re-estimated global flux with additional data to a figure of 0.43 PgC yr^{-1}. The largest fluxes come from the Amazon, from the Ganges Brahmaputra system and from Oceania (Figure 6.2). The IPCC 2013 report (Ciais et al. 2013; Cubasch et al. 2013) adopted a value of 0.9 PgC yr^{-1} for total carbon flux to the oceans (consistent with the estimates from Cole et al. (2007) and Aufdenkampe et al. (2011) discussed in Chapter 2), which is broken down to 0.2 PgC DOC, 0.1–0.4 PgC POC and 0.5 PgC inorganic carbon.

We can compare these figures that are derived from catchment outlets, with estimates of overall lateral organic carbon flux, based on estimates of surface erosion. Figure 6.3 shows average carbon flux values derived from a range of global surface

Figure 6.1 (a) Erosion on arable hillslope in East Sussex. Source: After Boardman 2013. Courtesy of John Boardman. (b) Severe rill erosion on cropland in Bedfordshire Source: Courtesy of John Quinton.

erosion estimates compiled by Kirkels et al. (2014). The estimated global mean value of 3.3 PgC yr^{-1} is an order of magnitude greater than estimated global POC yield to the oceans, which is consistent with geomorphological understanding that hillslope sediment delivery rates are relatively low. In the Belgian Loess belt, Wang et al. (2010) estimated that 50–80% of eroded soils from agricultural land are re-deposited as colluvium downslope, and Aufdenkampe et al. (2011) suggest that sediment delivery rates are typically in the range of 5–25%. Intervening storage in depositional landforms such as footslope deposits, alluvial and colluvial fans, and river floodplains mean that the majority of organic matter eroded from hillslope systems is at some stage returned to hillslope or floodplain storage.

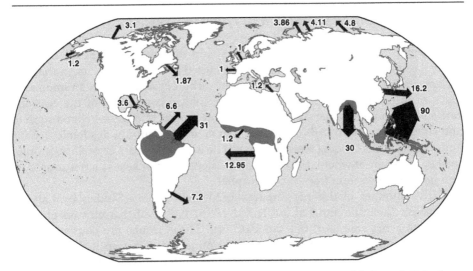

Figure 6.2   Global organic carbon flux to the ocean from major world rivers. Annual discharges in MtC yr⁻¹. Source: After Schlünz and Schneider 2000. Reproduced with permission of Springer Nature.

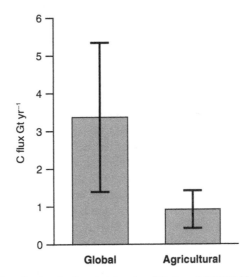

Figure 6.3   Estimated lateral carbon flux based on data from Kirkels et al. (2014) Table 4 (mid point estimates). Error bars are one standard deviation n = 7 (global); n = 6 (agricultural). Source: Martin Evans.

Consequently, whilst the fluvial carbon flux to the oceans is a significant component of terrestrial carbon cycling (Battin et al. 2009), mobilisation and storage of organic carbon on hillslopes is a major and understudied element of the system. Aufdenkampe et al. (2011) suggest that 'an order of magnitude more C is buried annually in stable continental deposits than is buried in the world's oceans', p. 57.

Sources and Sinks of Carbon in Hillslope Systems

The magnitude of erosional carbon fluxes, and the importance of the hillslope system, as both a source and store of remobilised organic matter, mean that unpicking the role of erosion in the terrestrial carbon cycle has been an important focus of research. The functional and geographical complexity of the erosion–carbon system means that quantifying the net impact of erosion-related sources and sinks of terrestrial carbon has been an area of significant academic debate, and remains a challenging research focus. The ongoing debate crystallises around the key question of whether accelerated hillslope erosion acts as a net source to, or a net sink of, carbon from the atmosphere

Early estimates of this net impact varied considerably. Studies by Lal and co-workers (Jacinthe and Lal 2001; Lal 1995) indicated that erosion of croplands was a net source of 0.4–0.6 PgC yr^{-1}. These results are consistent with the idea that soil organic matter is rapidly mineralised in transit. Jacinthe and Lal argue that the preferential detachment of lighter soil particles is associated with transport of more labile components of soil organic carbon, and that this carbon is rapidly oxidised and lost to the atmosphere. In addition, they suggest that carbon which has been physically protected from mineralisation in soil aggregates is exposed, as these aggregates are broken down in transport. Jacinthe and Lal quote potential carbon mineralisation rates of 20–70% associated with these processes. In contrast, Stallard (1998) focused on the storage of eroded carbon across a range of terrestrial sinks, including lake sediments, colluvial and alluvial sediments. Global rates of carbon storage (including all physical erosion not simply cropland) were estimated, with the net impact of soil erosion emerging as a carbon sink of 0.6–1.5 GtC yr^{-1}. Key to the conceptualisation of Stallard, is the notion of dynamic replacement (Harden et al. 1999). Preservation of re-deposited organic carbon in terrestrial stores can only be an ongoing net carbon sink from the atmosphere if the soil organic matter lost from erosional sites is replaced by new carbon, that is photosynthetically fixed from the atmosphere.

The divergence of conclusions in these early estimates of the impact of erosion has been replicated in subsequent work, which has concluded that the net impact of erosion on terrestrial carbon balance is either a source (Ito 2007; Lal 2003, 2004, 2008) or a sink (Berhe et al. 2007; Liu et al. 2003; McCarty and Ritchie 2002; Smith et al. 2001; Van Oost et al. 2007).

There is broad agreement in this literature as to the key processes which control the interaction of soil erosion and carbon cycling. Carbon storage is promoted by high rates of carbon deposition and carbon preservation in depositional areas. If this is supported by continued carbon accumulation at erosional sites (net primary productivity > respiration), then the landscape system will tend to accumulate carbon in response to erosional events. Larger sink estimates have assumed that this 'dynamic replacement' represents 100% of the eroded carbon, whereas lower sink estimates, such as those of Van Oost et al. (2007), are based on data which suggests that only the active soil carbon pool (circa 25% of the total) is replaced. Critical here is the impact of erosion on NPP at erosional sites and the role of soil improvement in supporting continued carbon accumulation at degraded sites in agricultural systems. Erosion also has the potential to reduce net carbon storage, because enhanced mineralisation of organic carbon occurs during the process of erosion as labile carbon is exposed and mobilised. The larger sink estimates, such as those from Jacinthe and Lal (2001), assume that circa 20% of SOC is lost in transit through this mechanism. The status of a landscape as a source or a sink is controlled by the relative magnitude of these effects, and can be assessed by comparing estimates of total landscape carbon storage for an eroding system against the balance of NPP and respiration for stable systems.

Doetterl et al. (2016) provide an excellent summary of the range of values for the erosional carbon balances which have been proposed (Figure 6.4), and argue that part of the explanation for the observed variation lies in the nature of the systems and sediments considered by varying studies, which is in turn influenced strongly by the disciplinary background of the authors. Doetterl et al. imply that

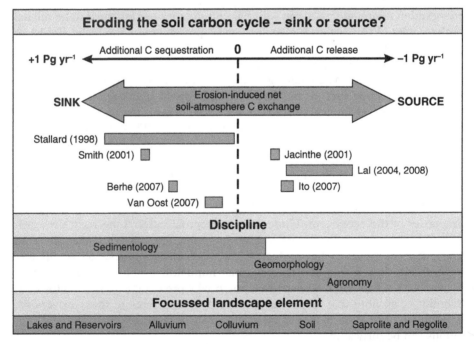

Figure 6.4   This plot positions key papers in the debate around whether erosion is a source or sink in a space which indicates both their disciplinary origins and the landscape components they study. It demonstrates that the study design and selection of time and space scales are fundamental considerations for studies of carbon landsystems. Source: After Doetterl et al. 2016. Reproduced with permission of Elsevier.

studies which report a link between carbon sinks and erosion are predominantly those which take a longer-term sedimentological approach to the problem by measuring long-term carbon sequestration rates. Inherent in this analysis, is a recognition of the complexity of the system and therefore of the sampling challenges associated with attempts to provide a reliable global estimate for these values. The importance of understanding mechanisms of carbon decomposition, of assessing carbon fluxes at appropriate time and space scales, and of considering landscape position as a way to deal with this complexity, are considered below.

Controls on Carbon Transformations in Eroding Systems

Conventional approaches to conceptualising soil organic matter consider a range of carbon pools within the soil system, which are differentiated by potential decomposition rates, linked to the molecular form of the carbon. For example, the widely used Century soil model (Parton 1996) incorporates an active carbon pool with turnover times of months to years, a slow pool with turnover times of up to 50 years and a passive pool with turnover times of 400–2000 years. Long-term preservation of organic matter in soils has commonly been assumed to be related to the potential for certain molecular forms (particularly lignins and lipids) to resist biological decomposition (Stevenson 1994), whilst more labile forms of carbon such as carbohydrates are rapidly recycled. More recently however, this paradigm has begun to be challenged. Studies such as work by Marschner et al. (2008) have argued that in the absence of additional stabilisation mechanisms such as sorption or physical protection, very little organic matter persists in sedimentary environments for periods in excess of 50 years. They showed that in the absence of physical protection, no class of organic compounds appeared to be more resistant than bulk organic matter.

On the basis of findings such as these, Schmidt et al. (2011) have argued that SOM persistence is not primarily a function of molecular structure, but is in fact controlled by environmental and biological controls, which can be regarded as properties of the ecosystem. Under this conceptualisation, long-term preservation of soil organic matter is driven not by molecular characteristics inherited from litter composition, but by factors such as hydrophobicity, soil acidity and mineral sorption. These control microbial populations and substrate variability, and vary across the landscape in a potentially predictable manner. However, Schmidt et al. note that responses of decomposition rates to these drivers are typically non-linear, so that simple correlations with average environmental conditions can be misleading.

On a global scale, Hemingway et al. (2019) have shown that changes in activation energy in a dataset of radiocarbon dated soils and sediments are consistent with the formation of organo-mineral bonds over time and inconsistent with selective preservation of recalcitrant material in older samples. This shift in understanding of organic matter preservation in the landscape has two important implications, which require an integration of geomorphological understanding into assessments of landscape scale budgets of organic matter. The first implication is that carbon preservation potential will vary in space, as a function of changing sediment character, but also of moisture and redox conditions, which are strongly associated with topographic variation. The second is that variation in carbon stocks within sedimentary stores of carbon may not typically reflect predictable decline in carbon concentration, for example the shift in quality towards more recalcitrant species with depth. Instead, variation may reflect the sedimentary history of a site, so that vertical variations in sediment character and long-term changes in sediment dynamics, may play a significant role in carbon preservation.

Spatial and Temporal Patterns of Organic Carbon Transformation

Topographic Variation

Environmental controls on microbial capacity to mineralise organic matter such as soil moisture and redox conditions are likely to vary with topographic position (Swift et al. 1979). Footslope positions and water accumulating sites have higher mean water tables generating lateral flow (Anderson and Burt 1978) and may develop reducing conditions during wet periods. Similarly, topographic position is a key control on soil erosion rates, with erosion focussed on steeper and longer slopes and deposition occurring in low-angled footslope areas. Topographic position is simple to compute using terrain analysis tools, and so offers a potentially powerful method to extrapolate carbon budgets to the landscape scale. The most comprehensive investigation of topographic control on carbon sequestration in eroding agricultural systems is work by Berhe and co-authors (e.g., Berhe 2012; Berhe et al. 2012). Berhe et al. (2012) assessed SOM quality in a range of topographic settings (Figure 6.5) and demonstrated that SOM varied between eroding sites (more transformed) and depositional sites (better preserved). The dominant controls on SOM preservation however were enhanced complexation with mineral matter and greater physical protection of SOM within aggregates in footslope depositional settings. These sites also had the oldest carbon (as determined by ^{14}C dating: Figure 6.5), indicative of longer-term preservation of carbon in these depositional settings. The importance of physical protection of organic matter at these sites was emphasised by the fact that direct measurement of decomposition rates for freshly added OM in depositional settings indicated much faster decomposition for shallow buried material, than *in situ* at eroding sites.

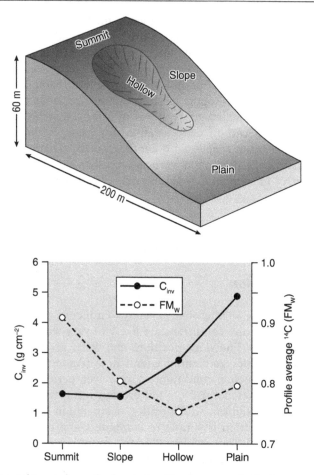

Figure 6.5   Topographic controls on soil carbon stock and the average age of soil carbon across the hillslope profile. Source: After Berhe et al. 2012. Reproduced with permission of Wiley.

Similar results have been derived from more direct measurement of decomposition rates based on ^{14}C aging of respired CO_2 along an elevational transect in the Sierra Nevada (Wang et al. 2010). Higher soil moisture contents (beyond a threshold of 14–25% water content) inhibited rates of SOM decomposition, although temperature controls on decomposition meant that along a significant elevational transect, the maximum decay rates were at the low elevation sites. At the landscape scale therefore, both topographic controls on wetness, and altitudinal control of temperature, are significant controls on rates of SOM decomposition.

Measuring Carbon Sequestration and Mineralisation at Appropriate Scales

The importance of scale and the magnitude of intermediate deposition, in determining reasonable transport losses for organic carbon in eroding systems, was recognised early in the study of erosional carbon losses. Van Noordwijk et al. (1997) suggested that erosional losses should scale at a power of 1.6 of the length scale of study, rather than the square area. Whilst this is an appropriate measure at the scale of a field, van Noordwijk et al. also argue that this scaling is applicable to landscapes, so that the proportion of eroded soil deposited in an erosion plot is similar to the proportion deposited at the landscape scale. This is justified with reference to Australian data relating catchment area to specific sediment yield, which shows decline over scales of 0.01 to 10 000 km^2 (McLaughlin et al. 1992). This scaling has also been demonstrated to be a widespread pattern globally (Church et al. 1989; Figure 6.6). However, Van Noordwijk et al. note that this 'fractal dimension' of sediment is likely to vary with local variations in relief and surface roughness (and this is evident in the empirical data they cite). While the relation may be reasonable for specified lithologies and tectonic settings, variations in landscape hypsometry and deviations in distribution of slope in space mean that sediment delivery ratios – the proportion of eroded sediment which leaves a catchment and is not deposited in intervening depositional sites – vary significantly in space (Figure 6.7).

Equally important in understanding spatial patterns of erosion and sedimentation, and by implication patterns of organic matter flux through catchment

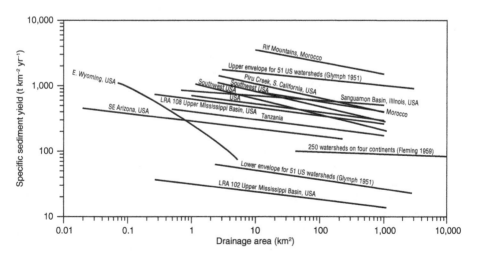

Figure 6.6   Variation of specific sediment yield with catchment area. Source: After Church et al. 1989. © Canadian Science Publishing or its licensors.

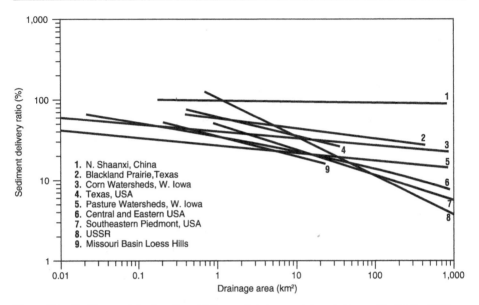

Figure 6.7   Significant variation in sediment delivery ratio between catchments. Source: After Walling 1983. Reproduced with permission of Elsevier.

systems, is landscape history. Whereas it might be reasonable to assume that under conditions of constant climatic forcing and consistent hypsometry that sediment delivery ratios would be consistent between landscapes, in fact the historic exposure of landscapes to different process regimes can lead to distinct landscape configurations, which modify spatial patterns of sediment flux. For example, Church and Slaymaker (1989) demonstrated that in British Columbia, patterns of specific sediment yield diverge from the conventional pattern, with elevated specific sediment yields in catchments of up to 10 000 km². This pattern is ascribed to a disequilibrium effect of deglaciation. Large quantities of sediment released as glaciers melted have been redeposited along river courses and are being reworked through the fluvial system, elevating local sediment yields.

At a smaller scale, episodic cutting and filling in fluvial systems (Kirkels et al. 2014; Schumm 1973) has a similar effect of complicating spatial patterns of organic sediment flux. The classic Trimble (1983) study of elevated sediment flux in response to upland agriculture in Wisconsin demonstrated very rapid alluviation in tributary valleys, whilst sediment flux from the wider catchment remained constant. These sediments were then remobilised further down the system, after the cessation of accelerated upland erosion. In this example, the sediment delivery ratio at the catchment outlet varies in inverse proportion to overall erosion, whereas at the head of the main valley this impact is dampened.

Assessment of carbon sequestration is also highly sensitive to temporal scale of measurement. Lateral fluxes of carbon in common with all sedimentary systems can be highly episodic. Frith et al. (2018) report lacustrine records of accumulation of biospheric carbon from forested steeplands in New Zealand. Pulses of sediment relating to four earthquake events in the past 1100 years account for 24% of the core record length but account for 43% of the preserved biospheric carbon. These results emphasise both the importance of large rare events in driving connectivity in these sediment systems, but also the potential limitations of short-term monitoring of terrestrial carbon cycling.

What these examples demonstrate is that at the landscape scale, patterns of erosion and of deposition, and consequently patterns of lateral carbon flux, storage and decomposition are unlikely to remain scale invariant in topographically complex landscape systems. Much of the measurement which has informed the discussion of erosion and the terrestrial carbon cycle has taken place on relatively small plots, often in lowland agricultural settings with relatively low relief. Variations in sediment flux and sediment delivery ratio in space and time make simple scaling of plot measurements challenging. An understanding of landscape sediment budgets is necessary to define appropriate spatial scales for the measurement of carbon sequestration and mobilisation within the geomorphological system. Of course, in order to understand the fate of sedimentary organic carbon, it is necessary to understand not just where it is being stored and in what quantities, but also for how long, since carbon mineralisation is a time-dependent process.

Carbon Transformation at Erosional Sites

At erosional sites, losses of particulate carbon from the soil occur at the timescale of individual erosional events driven by the return period of significant rainfall events. However, key to understanding the carbon balance of these sites are the timescales of the ecosystem response to ongoing erosion. Dynamic replacement of eroded carbon is a key process here. Harden et al. (1999) describe a site in Mississippi where all of the historic soil carbon was lost during 127 years of intensive farming, whilst in the 50 years since 1950, 30% of this carbon loss has been replaced. Harden et al. argue that with the intensive application of fertilisers and consequent increases in primary productivity, these systems have become net carbon sinks. This is consistent with the observation by McLauchlan (2006) that SOC accumulation in agricultural systems is primarily determined by carbon inputs from crop residue. Dungait et al. (2013) directly assessed dynamic replacement at eroding sites, through the use of isotopically labelled maize, and demonstrated that the labelled litter contributed most to soil microbial biomass carbon in topslopes (eroding) sites and least in footslopes (depositional sites).

Carbon sequestration at disturbed sites may also be enhanced through exposure of fresh mineral binding sites, which can stabilise newly produced SOM, although at some sites erosional losses may inhibit primary productivity (Lal 2004). Timescales for disturbed sites to reach a new equilibrium SOM concentration are commonly reported to be decades up to 75 years (McLauchlan et al. 2006; West and Post 2002), so that equilibrium rates of dynamic replacement in response to onset of erosion can be expected to develop within this timescale.

Carbon Transformation During Sediment Transport

Acceleration of SOM mineralisation due to erosion (Jacinthe and Lal 2001; Lal 2003) can occur due to oxidation of eroded SOM during erosional episodes, and as a result of modification of eroded material which is transported to depositional sites. This can occur because of physical disruption of soil aggregates, or by the movement of organic matter from local environments which favour preservation (acidic, low oxygen), to sites which favour decomposition. The key processes are then physical erosion, which occurs at the short timescales (minutes to hours) of erosional events, decomposition of organic matter, which may be mediated either by microbial processes (Davidson and Janssens 2006) (timescales of days to weeks), or by the action of UV light, particularly on dissolved carbon (Worrall et al. 2012) (timescales of minutes to days).

Direct measurement of carbon losses in transport are lacking in the literature. Reported rates of mineralisation of eroded material vary from negligible to 20% carbon loss (Kirkels et al. 2014), but typically assess not only within-event losses, but also losses from recently deposited material. For example, Van Hemelryck et al. (2011) is one of the few studies to directly assess CO_2 flux from eroded material, measuring carbon balance at a site of fresh field deposition for a period of 112 days. This study supported the idea that eroded material was subject to accelerated mineralisation over a period of a few weeks. However, the total carbon loss associated with this period of elevated respiration rates was only 1.6% of the soil carbon estimated to be mobilised during the erosive event.

Quinton et al. (2010) have argued that the relative importance of preferential oxidation of eroded material is strongly linked to spatial scale. At the scale of in-field erosion and deposition, timescales of transportation are very short, so that losses in transit are likely to be small because the period of transport is shorter than the timescales required for significant microbial decomposition to occur. However, at larger spatial scales, losses can be much greater because of longer travel times, rapid losses of DOC from fluvial systems and the potential for longer-term oxidation of organic matter in the fluvial system (Chapter 4).

Carbon Transformation at Depositional Sites

Relevant timescales for understanding carbon turnover at depositional sites can be very long (centuries or longer). The understanding of depositional landforms as sites of carbon sequestration assumes that stabilisation of organic matter (through mineral adsorption or because the physical environment is not conducive to microbial decomposition) occurs relatively rapidly limiting *in situ* mineralisation of deposited organic carbon. In accumulating sediment systems, it is assumed that more recalcitrant organic matter is preserved because it is not a favoured substrate for microbial decomposition. Consequently, SOM age is expected to increase with depth and the proportion of recalcitrant materials such as lignins and lipids to increase.

The proportion of deposited organic carbon preserved in the sediment sequence is described as the carbon burial efficiency. In agricultural soil systems in central Belgium, Wang et al. (2010) have shown carbon burial efficiency reached a constant value after 1000–1500 years. In this study, incubation experiments showed rapid decomposition of labile material and stable OM composition below the upper layers of the soil, consistent with the preservation of recalcitrant species. These results are consistent with conventional understanding of carbon cycling in depositional settings. However, as noted above, the ideas of Marschner et al. (2008) and Schmidt et al. (2011) suggest that there is potential for decomposition of all forms of organic matter in relatively short timescales under suitable biophysical conditions.

In stable depositional settings, carbon sequestration requires preservation of OM at timescales in excess of 50 years, which Marschner et al. suggest is sufficient for decomposition of unprotected OM in soils. Complex patterns of carbon preservation have been observed, linked to variation in microbial niches within sedimentary systems. Ekschmitt et al. (2008) argue that habitat constraints on the activity of microbial decomposers are a key control on carbon preservation. They showed that high levels of specialisation of decomposer organisms in relation to carbon substrate means that the measured age of stored carbon and the residence time of organic matter in the system does not necessarily coincide with the relative lability/recalcitrance of the organic compounds. Controls on decomposition can be relatively large scale, for example anaerobic conditions below the water table limit fungal decomposition of lignins, as the key enzymes require oxygen. They can also vary at the microscale, independent of burial depth. For example, bacteria can be transport limited, so that small-scale variability in the environment such as microscale hydrophobicity may limit access to organic matter. Essentially, Ekschmitt et al. argue that organic matter preservation occurs in environmental niches which are favoured in footslope and depositional geomorphic settings. They state that 'evolutionary selection towards specialised substrate utilisation has provided the decomposer community with the chemical and physical tools to degrade virtually all kinds of soil organic components on the long term', p. 33.

Similar findings have been presented by Sagova-Mareckova et al. (2016), who showed that bacterial communities in deep colluvial soils reflect soil carbon content and local chemical conditions, so that the potential for carbon preservation is best conceptualised as 'patches' of preservation rather than gradients of preservation. A logical consequence of these patterns is that organic matter can be lost from depth. Jörgensen et al. (2002) recorded declines in microbial biomass and carbon down profile through a deep buried chernozem and showed that a large part of original SOM was lost from this buried organic layer. Kramer and Gleixner (2008) report declines in microbial biomass with depth and increasing substrate specialisation. OM decomposing organisms occur at depth, but Kramer and Gleixner emphasise the importance of priming of the microbial system by labile carbon transported from the surface in triggering deep OM decomposition.

The potential for decomposition of organic matter at depth and for microbial decomposition of 'recalcitrant' materials in suitable conditions, has a potentially important implication in terms of the overall carbon budget. Anaerobic decomposition pathways (more common in deeper soils and in wetter footslopes) release gaseous carbon in the form of methane. Because methane has a much higher greenhouse warming potential than CO_2, there is the possibility that even if the carbon balance of eroding landscapes indicates a net carbon sink, the greenhouse gas potential could infer net warming potential. Lal (2008) argues that this is a potentially important mechanism and represents a key unknown in assessing the overall impact of hillslope erosion. Deposition of eroded material on the surface of valley floor wetlands with characteristically high water tables will tend to favour methane production, whereas hillslope deposition at sites such as hedgerows or behind soil conservation measures are less likely to be saturated and may promote mineralisation of carbon to CO_2.

Carbon preservation in deep soils is empirically demonstrated through enhanced carbon concentrations in footslope soils and through the presence of aged ^{14}C in soil profiles (Berhe et al. 2012). Note that Berhe and Kleber (2013) argue that dilution of old SOM with young eroded SOM in depositional areas can lead to young ^{14}C ages, but still imply overall carbon storage. Chaopricha and Marín-Spiotta (2014) reviewed the literature on carbon burial in deep soils and argue that soil burial stabilises SOC and leads to the preservation of carbon at timescales of over 1 ka.

The evidence discussed here is consistent with the assertion that carbon sequestration in depositional landforms can lead to carbon storage at millennial timescales. However, the evidence also indicates that decomposition may be an ongoing process even in deep soils, and that ecosystem changes which modify the microbial niches within the soil and sediment structure, can lead to renewed mineralisation of organic matter in previously stable systems.

Timescales for Measurement of Hillslope Carbon Transformations

Based on the discussion above, Figure 6.8 identifies key processes and dominant timescales for carbon transformation across the hillslope system, from sites of hillslope erosion, to footslope depositional zones. The variation in the dominant timescale between hours to minutes for hillslope transport, and up to millennia

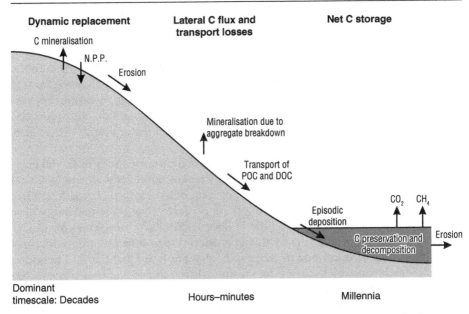

Figure 6.8   Key carbon transformation processes on hillslope systems and characteristic timescales. Source: Martin Evans.

for carbon transformation in footslope zones, means that assessing carbon sequestration rates at a site is dependent on both the temporal and spatial scales of measurement. A full appreciation of these systems also requires a diverse range of methods, from direct measurement of contemporary gaseous carbon flux, through to measurement of sedimentary carbon storage in depositional zones. In addition to assessment of rates of storage and transformation of sedimentary carbon, a complete analysis also requires understanding of the long-term geomorphic stability of depositional landforms where carbon is stored, since this defines the upper limit of carbon residence times within the system.

Landform Stability and Long-term Carbon Storage

Chaopricha and Marín-Spiotta (2014) note that whilst deep burial of carbon can lead to carbon storage, disturbance of these soils can lead to rapid decomposition at timescales of tens of years, due to oxygenation and wetting of sediments. Disturbances considered in the review include climate change and human intervention, but the potential for decomposition of organic matter at depth and the possibility of microbial decomposition of 'recalcitrant' materials in suitable conditions, also means that the longer-term sedimentary dynamics of the system may be a key control on timescales of carbon preservation. Where there is long-term stability in depositional sites, then net carbon sequestration can occur. However, when sites are disturbed, there is prospective release of the stored carbon to the atmosphere. This may result from *in situ* mineralisation of carbon in disturbed or aerated soils, or it may represent complete loss of carbon through erosion to the fluvial system, and subsequent oxidation of carbon within the fluvial system (see Chapter 7). Therefore, a complete understanding of controls on carbon seques-tration in hillslope depositional landforms, such as colluvial fans and footslope deposits, requires an understanding of the age of such features and of their per-sistence in the landscape.

Lewin et al. (2005) present radiocarbon dates for a wide range of fluvial envi-ronments in the UK and note an absence of early Holocene alluvial/colluvial fans, which they regard as unusual in the context of a paraglacial landscape. They infer that earlier fans deposited during periods of deglacial slope instability have been eroded and reworked during the Holocene. Similarly, timescales for the re-working of valley side deposits by fluvial erosion due to meander migration may be on the order of hundreds to thousands of years (see Chapter 7). Therefore, the timescales over which carbon loss from colluvial deposition may be expected to occur, overlap with the probable sediment residence times in some colluvial contexts. For this reason, understanding the connectivity of hillslope sediment storage to the wider sediment cascade is important, since it is a potentially critical control on carbon residence times.

Landscape Connectivity

The importance of connectivity between landscape sub-systems as a control on sediment flux to a catchment outlet has long been recognised (Brunsden and Thornes 1979; Harvey 2001). This has typically been conceptualised in the con-text of the sediment budget, so that connectivity is understood as a key con-trol on material flux between distinct sediment stores. These sediment stores are either individual depositional landforms or characteristic suites of depositional landforms within a landscape type. More recently, Bracken et al. (2015) have argued that connectivity should be conceptualised as a continuum of material flux between sources and stores, controlled by sediment detachment processes and characteristic travel distances for sediment. Bracken et al. note that the contingent arrangement in time and space of sediment detachment and trans-port processes becomes more variable with increasing spatial scale. Therefore, in relation to carbon storage, the relative importance of understanding sedi-ment dynamics alongside biological controls on carbon cycling increases with catchment size.

It is important to recognise that landscape connectivity is not a fixed param-eter, but can vary significantly over time. A good example comes from the work of Houben et al. (2013) working in Loess basins in Germany, who demonstrated a delay on the order of 103 years between the onset of footslope alluviation in response to Neolithic agriculture and subsequent aggradation of floodplains. They argue that 'changes in alluvial sedimentation are strongly influenced by the spatial organisation as well as the functional (hydrosedimentary) connectivity between upcascade components, and, thus, largely independent from processes of sediment redistribution on hillslopes', p. 551. This is an important point. In the context of their study, the changes in functional connectivity are driven by changes in land use, but their argument also holds for changes in alluvial behaviour driven by hydroclimatic change, which may modify rates of reworking of hillslope material to the fluvial system.

A potential way forward in integrating understanding of landform dynamics with consideration of carbon storage in the landscape has been explored by Torres et al. (2017). This study integrates a meandering river model with a POC cycling model, which is validated against known ^{14}C ages of floodplain organic matter. The model, based on the idea that the stochastic nature of fluvial processes leads to rivers that preferentially erode young deposits, shows that stored material is older than expected and therefore that a proper understanding of the controls on the sedimentary carbon cycle include sediment dynamics, as well as the turnover of modern material. A similar modelling approach, which considered connectivity of the hillslope-channel system by assessing changes in channel position relative to valley edge footslope deposits, could provide the basis for an assessment of the impact of hillslope erosion on carbon cycling, which explicitly considers landscape scale and connectivity as controls.

The overall magnitude of hillslope and floodplain carbon storage has been assessed in an important study of the Rhine basin by Hoffmann et al. (2013b) through compilation of data from over 1500 sediment sequences in hillslope deposits and river floodplains (Figure 6.9). The findings of this study, that hillslope carbon storage exceeds floodplain storage at spatial scales up to 10^5 km^2, is consistent with relatively low connectivity across the large floodplain of the Rhine system. In this system, Hoffmann et al. (2013b) assess that hillslope and floodplain carbon storage are two orders of magnitude greater than that in lakes and reservoirs. The magnitude of the surface store is sufficiently large that it exceeds the greater carbon burial efficiency of lake and reservoir systems, so that the surface sediment carbon stores are dominant sites of both carbon storage and carbon turnover.

The Hillslope Carbon Land System

Figure 6.10 and Table 6.1 present a carbon landsystem model of hillslopes, based on the current understanding of key controls on hillslope carbon cycling discussed in this chapter. The focus of this landsystem is on agricultural systems and

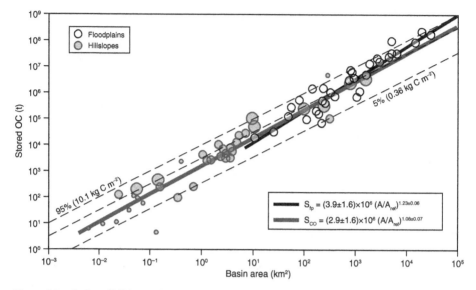

Figure 6.9   Scaling of hillslope and floodplain carbon storage across five orders of magnitude in the Rhine basin. Source: After Hoffmann et al. 2013b. Reproduced with permission of Wiley.

on hillslope erosion processes characteristic of these systems, including tillage erosion, erosion by rainsplash and un-concentrated flow, and gully erosion. A complete hillslope model would incorporate steepland processes such as landsliding, which have been characterised in the headwater model presented in Chapter 5. Table 6.1 considers the key processes of carbon cycling operating across the hillslope catena, the timescales at which they operate, and the implications for the hillslope scale carbon balance.

We can conceptualise the carbon balance of the hillslope as the sum of the carbon balance of three slope segments as in Equation 6.1:

Equation 6.1: $H = U + M + L$

where H is the net carbon balance of the hillslope system and U, M and L represent the local carbon balance of the upper, mid and lower slope segments. For each hillslope segment the local carbon balance can be expressed as the product of Equation 6.2:

Equation 6.2: $NPP - R - L$

where NPP is net primary productivity, R is heterotrophic soil respiration and L (Lateral losses) is removal of SOM by downslope erosion.

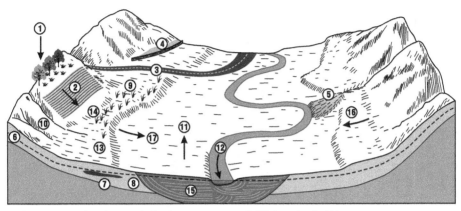

1. Erosion and dynamic replacement of soil carbon
2. Tillage erosion on midslopes.
3. Enhanced connectivity of OM flux through sites of concentrated flow such as roadways.
4. Midslope storage of OM behind anthropogenic features such as field boundaries.
5. Reactivation of depositional fan deposits by fluvial erosion.
6. Hillslope water table.
7. Buried soils in footslope deposits.
8. Deep soils in footslope positions, and C storage due to high water tables.
9. OM storage in colluvial fans.
10. Hillslope gully erosion.
11. Oxidation of OM stored in floodplain deposits.
12. Fluvial carbon flux.
13. Mineralization of OM in footslope deposits.
14. Footslope storage of deposited OM.
15. Floodplain deposits.
16. Erosional C flux through rainsplash and unconcentrated flow.
17. DOC flux from decomposition of footslope deposits.

Figure 6.10   A hillslope carbon landsystem. 1) Erosion and dynamic replacement of soil carbon; 2) Tillage erosion on midslopes; 3) Enhanced connectivity of OM flux through sites of concentrated flow such as roadways; 4) Midslope storage of OM behind anthropogenic features such as field boundaries; 5) Reactivation of depositional fan deposits by fluvial erosion; 6) Hillslope water table; 7) Buried soils in footslope deposits; 8) Deep soils in footslope positions, and C storage due to high water tables; 9) OM storage in colluvial fans; 10) Hillslope gully erosion; 11) Oxidation of OM stored in floodplain deposits; 12) Fluvial carbon flux; 13) Mineralization of OM in footslope deposits; 14) Footslope storage of deposited OM; 15) Floodplain deposits; 16) Erosional C flux through rainsplash and unconcentrated flow; and 17) DOC flux from decomposition of footslope deposits. Source: Martin Evans.

The land system model demonstrates the spatial complexity of the land system and emphasises the importance of hillslope connectivity in controlling lateral carbon losses from the lower slopes. Whether any given hillslope catena or hillslope segment is a source or a sink of carbon is strongly determined by local configuration of the hillslope system. In agricultural systems at the hillslope scale, this is a function of anthropogenic activity on the hillslopes (features such as roads, walls, tillage techniques, configuration of arable and pastoral activity in space). The potential degree of spatial variability means that new empirical work on understanding hillslope carbon cycling processes must be carefully contextualised. Although there is still some controversy in the literature, the best empirical evidence suggests that footslope deposits are sites of carbon storage, so that $L > U$. Thus, at the scale of the hillslope, soil erosion is a net sink of carbon,

Table 6.1 Processes of carbon transformation on hillslopes underpinning the carbon landsystem model presented in Figure 6.10.

	Key processes	Relevant timescales	Carbon budget
Upper	• Loss of SOM through erosion (rainsplash erosion, tillage erosion and rill/gully erosion) • Fixation of carbon from the atmosphere by NPP – NPP driven by nutrient status with an important role for fertiliser in agricultural systems. • Mineralisation of carbon from subsoil • Adsorption of OM to fresh mineral surfaces exposed by erosion	• Erosional events – hours • Equilibration of SOC content in response to changing vegetation or erosion – Decades	Lateral losses drive net carbon loss from upper slopes. The magnitude of the loss can be mitigated in natural systems by accelerated C fixation in young vegetation or by intensive fertilisation in agricultural systems
Mid	• Physical aggregate breakdown during erosion events • Storage of sedimentary carbon behind hillslope obstructions such as field boundaries	• Carbon transformations of mobile organic matter occurring during erosional events – minutes to hours	*In-situ* carbon balance may be in equilibrium with lateral inputs balanced by erosional outputs Mineralisation in transit limited by short timescales of events
Lower	• Episodic deposition of hillslope organic matter during erosional episodes • Microbial decomposition of OM in aerobic conditions to CO_2 and in anaerobic conditions producing CH_4 • Priming of the microbial decomposition of older OM deposit by fresh labile carbon • OM stabilisation by adsorption to fresh mineral surfaces • Reworking of footslope deposits through fluvial action or climatically driven incision	• Accelerated decomposition of material deposited during erosional events – weeks • Carbon storage and turnover at timescales of up to 1500 years • Physical reworking of stored carbon in footslope deposits to the fluvial system at timescales of centuries to millennia	Depositional sites are net carbon sinks with *in-situ* carbon accumulation in wet footslope sites supplemented by deposition of OM from hillslope deposition At longer timescales, erosion of footslope deposits may make these sites net sources of carbon to the fluvial system

as indicated by the work of Van Oost et al. (2007). However, key to understanding the role of erosion in the wider terrestrial carbon cycle is the mismatch in the characteristic timescales that control key processes of carbon loss and sequestration in the upper and lower parts of the hillslope system.

Footslope deposits have high water tables and relative chemical stability, due to mineral stabilisation of OM by adsorption and inhibition of microbial decomposition, either through the development of anaerobic conditions or physical isolation of OM within mineral aggregates. However, at timescales of centuries to millennia, landscape scale sediment dynamics dictate that footslopes experience physical instability. Erosion of footslope deposits may be driven by lateral shifts in river course, creating local coupling of the hillslope-channel system (e.g., Harvey 2001) or by incision of hillslope deposits driven by changes in land use or climate (e.g., Chiverrell et al. 2008). Shifts in the carbon balance of upper hillslope segments involve changes in the balance of lateral erosion and net primary productivity and have a characteristic time to reach a new equilibrium, which is in the order of decades, whereas changes in the export of OM from footslope deposits due to physical instability are likely to develop at longer timescales. What this means is that the carbon balance of upper and lower hillslope segments may be driven by processes operating at diverse timescales, so that changes in the time base for calculation of carbon budgets may significantly impact the estimated net impact of hillslope erosion on carbon cycling.

Hillslope Erosion and Terrestrial Carbon Cycling: A Research Agenda

Explaining and simplifying the diversity of global estimates of the impact of hillslope soil erosion requires both a clarification of the conceptual framing of the erosion–carbon sequestration debate and new empirical understanding of the magnitude of some key fluxes. Here, a series of crucial areas for further research are proposed, based on the understanding of hillslope carbon flux developed above.

Defining Spatial Scales

Perhaps the best estimate of erosion related carbon losses is the thoroughly validated modelling study of Van Oost et al. (2007). This is a true study of hillslope carbon balance, since it is based on upscaling findings from plot and field scale measurements of carbon stocks, to global cropland areas through the use of USLE (Universal Soil Loss Equation) soil erosion modelling. What this approach cannot assess however, is remobilisation of hillslope deposits. The distinction that Houben et al. (2013) draw between local deposition and remobilisation of sediments in a connected landscape is highly pertinent here and is reflected in the scaling of carbon storage in the Rhine (Hoffmann et al. 2013b). Defining appropriate spatial scales is key to interpreting empirical evidence on carbon storage. Further work on understanding spatial and temporal variability in the efficiency of hillslope-floodplain sediment connectivity, and developing models of these interactions, is required to link measurement of hillslope carbon stocks to estimates of catchment carbon loss from fluvial systems.

Defining Temporal Scales

Appropriate timescales for study of erosion related carbon fluxes are intimately linked to considerations of spatial scale. At the plot and field scale, the time taken for disturbed soils to reach new equilibrium SOM content represents an appropriate timescale of study, so that the 50 year averaging implicit in the use of Caesium 137 approaches to carbon stock estimation (Van Oost et al. 2007) is appropriate. However, relevant timescales for the connection of hillslope sediment stores to the fluvial system are on the order of millennia, so that the mobilisation of sediment pulses through catchment systems at the landscape scale occurs at these longer timescales (Church and Slaymaker 1989; Trimble 1983). Consequently, time and space scales may not scale linearly, since a shift from the hillslope to the catchment scale involves a shift of process domain. Carbon flux on hillslopes is driven by: sedimentary processes operating at timescales of hours to days (erosion and deposition events); microbial decomposition of organic matter operating at timescales of days to centuries (depending on site conditions and dominant metabolic pathways); and by timescales of vegetation response and SOM equilibration, which are on the order of decades. At the catchment and landscape scales, carbon flux is influenced by hillslope sediment delivery, but also by the geomorphological processes which control slope channel connectivity, operating at timescales of centuries to millennia. Superimposed on these patterns are further temporal complexities resulting from the non-stationarity of carbon erosion and deposition rates, linked to human activity and climate change (Kuhn et al. 2009). Disentangling the relevant timescales controlling carbon flux through hillslope systems requires further consideration of the age of carbon associated with the main stores and fluxes. Radiocarbon analysis of footslope storage, catchment sediment fluxes and the gaseous carbon flux from eroding and depositional sites are a promising way forward.

Quantifying Gaseous Fluxes

Estimates of carbon loss to the atmosphere associated with soil erosion have largely been derived from mass balance calculations, relating SOC storage at erosional and depositional sites, or estimated from laboratory studies. Calculation from stocks often involves calculating a small residual from the comparison of two large stocks, with the inherent risk that this involves. With some honourable exceptions (e.g., Van Hemelryck et al. 2011), there has been little focus on direct measurement of gaseous carbon fluxes. Direct gas flux measurements using laser spectroscopic approaches are now routine measurements and offer the potential to directly test assumptions of enhanced gaseous flux from eroded soil. There is also the potential to measure CH_4 flux alongside CO_2, in order to assess the greenhouse warming potential of changes in carbon cycling, as well as the impact on the simple carbon budget.

Understanding Microbial Controls on OM Decomposition

Understanding SOM persistence as an ecosystem property as proposed by Schmidt et al. (2011) focusses attention on the range of microbial niches that exist within stable and eroding soil systems. Much of the work in this area has focussed on analysing microbial biomass, or on assay of the enzymes which drive decomposition. Modern genomic approaches offer the potential to more routinely consider microbial populations within these niches, and the ways in which they are impacted by erosion and deposition of SOM. Critically, as Schmidt et al. note, relating ecosystem function to genomic characterisation of microbial communities requires an understanding of functional redundancy in these communities, so that approaches which monitor active gene expression through RNA analysis are required to unpick the processes of decomposition and key biophysical controls on these processes.

Integrating Ecological and Geomorphological Understanding into Modelling Approaches

As Doetterl et al. (2016) have noted, some of the variability in current estimates of the impact of erosion on carbon cycling derive from disciplinary differences in initial assumptions. A large part of the important work in this area has been done by soil scientists, which has produced a focus on agricultural systems, near-surface soil horizons and relatively short timescales. A more geomorphological perspective; for example, in the work of Hoffmann et al. (2013b) and Kuhn et al. (2009), requires attention to longer timescales, but introduces the further challenge of upscaling.

Ecological approaches to understanding carbon sequestration, focussing on the response of microbial communities to changes in their sedimentary niche, require an integration of short-term microbiological responses with longer-term evolution of the sedimentary system and of surface vegetation. At the global scale, much of this complexity may be disregarded in modelling approaches or averaged through empirical relations with simple biophysical drivers (e.g., Parton 1996). However, attempts to manipulate soil carbon storage in order to mitigate the impacts of anthropogenic carbon release to the atmosphere require a more complete process understanding.

The call for truly interdisciplinary approaches to this problem has been frequently made in the literature (Evans et al. 2013; Schmidt et al. 2011). The urgency of the necessity to answer the question of whether erosion is a net source or sink of carbon to the atmosphere has inevitably led to a focus on large-scale modelling work, which has made significant progress. However, it is argued here that to make further progress in this area requires a more complete understanding of process. The locus of new interdisciplinary work must be at the hillslope and small catchment scale, where a complete description of carbon cycling for a range of systems, including understanding of the geomorphological, macro scale ecological, and microbiological controls on carbon flux, and measurement of the particulate, dissolved and organic fluxes, has the potential to provide an integrated process understanding which can underpin future modelling work.

Conclusions

Stallard (1998), in early work on the relation between erosion and carbon cycling, identified the importance of understanding 'eco-geomorphology' as an approach to constraining the impact of erosion on terrestrial carbon cycling. This is a call which has recently been renewed by Kuhn et al. (2009) and Kirkels et al. (2014). As this book has argued for other elements of the sediment cascade, this requires an integration of geomorphological approaches to understanding landscape stability, and to quantifying organic sediment budgets at the landscape scale, with a process-based understanding of biological controls on organic matter fluxes between key sedimentary environments and the atmosphere.

The importance of topographic position and geomorphological context (both stratigraphic context and related to landform/carbon store stability) as controls on hillslope carbon budgets suggests that different landscape types may have characteristic impacts on the terrestrial carbon balance. Schmidt et al. (2011) argued that organic matter persistence was an ecosystem property, but through incorporation of the idea of characteristic suites of landforms, and relative landform stability, it is possible to extend this argument to suggest that the hillslope carbon budget is in fact an ecosystem property. It is argued that the notion of a carbon landsystem can form the basis of the eco-geomorphological approach, which Kirkels et al. (2014) and Kuhn et al. (2009) argue is required to fully address the question of whether hillslope systems are sources or sinks of organic carbon. The wide range of estimates of the global carbon budget linked to hillslope soil erosion is indicative of the complexity of the challenge, but it is argued here that an eco-geomorphological research agenda should focus on characterisation of carbon stores and fluxes for characteristic elements of the hillslope carbon cycle. Solving the source-sink question will require spatially distributed modelling of carbon flux, which incorporates the range of environmental, and physical and chemical controls on OM preservation in the landscape. The complex and non-linear nature of these controls suggest that rather than global modelling approaches, upscaling from well constrained carbon budgets for well-defined carbon landscape types is an optimal approach to constraining the global budget.

In terms of application, arguably more important than the precise definition of the global net carbon flux linked to hillslope erosion, is understanding of the processes – critical here is not the defining of a relatively small term in the global carbon budget, which is comprised of some source sites and some sink sites, but understanding what controls this variability. Hillslope systems are very significantly under human management, and perhaps in the Anthropocene, our focus should be on modification of management practices to maximise soil carbon storage, with positive impacts on agriculture as well as the carbon budget.

Chapter Seven
The Role of Floodplains in Terrestrial Carbon Cycling

Introduction

River floodplains are fundamental building blocks of the natural landscape. Globally, floodplains are estimated to cover in excess of 2 million km² (Tockner and Stanford 2002). As fluvial systems have increasingly been recognised as dynamic components of the terrestrial carbon cycle (Battin et al. 2009), the importance of understanding the role of floodplain systems in the fluvial carbon cycle has become clear. Floodplains are commonly defined as the low-lying land, prone to flooding, adjacent to river channels (Figure 7.1). However, this definition casts the river floodplain as a passive backdrop, across which biological and biogeochemical processes control fluxes of carbon to and from the floodplain. In fact, a long history of geomorphological research has developed an understanding of floodplains as dynamic components of the fluvial sediment system, which are formed and reformed across a range of timescales. Erosion of, and deposition onto, floodplains are key controls on the flux of particulate carbon to and from floodplain settings. A supply of fresh material to floodplain surfaces means that the scale and timing of deposition have the potential to control the parent material of floodplain soils, exerting a major control over floodplain soil carbon processes and carbon flux.

Hoffmann et al. (2013a) have demonstrated that floodplains have the potential to be important stores of carbon at Holocene timescales, so that analysis of these systems is required as part of understanding the terrestrial carbon cycle. They argue that further understanding of the systems requires integration of biogeochemical and geomorphological knowledge and approaches. Research

Geomorphology and the Carbon Cycle, First Edition. Martin Evans.
© 2022 Royal Geographical Society (with the Institute of British Geographers). Published 2022 by John Wiley & Sons Ltd.

Figure 7.1   Floodplain of the River Aln, Northumberland. Ox bow lakes and cut offs indicate significant lateral instability and the significant depth of fine-grained alluvium is apparent in the channel banks in the centre of the image. Note also large organic debris on the point bar and floodplain surface on the right-hand side of the image. Source: Martin Evans.

on biological carbon cycling in floodplain systems has typically paid insufficient attention to the dynamism of floodplain sediment systems, but equally many geomorphological considerations of floodplain carbon cycling have focussed on sedimentary carbon, and not sufficiently considered interactions of the sedimentary and biological systems.

Quantification of rates of sedimentary carbon storage in floodplain systems is an approach which draws on many of the geomorphologist's core skills relating to the construction and quantification of sediment budgets (Brown et al. 2009). Such analysis defines the role of floodplain systems in sequestering carbon from the atmosphere and makes a significant contribution to the understanding of fluvial carbon budgets. However, this approach focuses on the residual term (i.e., carbon storage resulting from the balance of carbon fixation and mineralisation) of the floodplain carbon budget (Figure 7.2). Fluvial organic matter is stored on floodplains at timescales which far exceed the travel time of river waters, so that floodplain systems are potentially hotspots of carbon turnover, as well as zones of carbon storage.

A complete understanding of the floodplain carbon budget has the potential to identify floodplain contexts where carbon turnover is important. It is important to

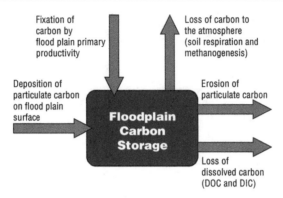

Figure 7.2 The floodplain carbon budget. Source: Martin Evans.

recognise that these locations can also be loci of carbon storage. Integrating a bio-geochemical approach into the geomorphological analysis of floodplains has the potential to explain the role of these systems in both carbon storage and carbon turnover. This chapter aims to summarise a thorough review of the relevant literatures, to bring together geomorphological and biogeochemical understanding of floodplain systems, to develop a conceptual model of floodplain carbon cycling, and to analyse controls on the variation of floodplain function in time and space.

Floodplain Systems

Sediment Storage on Floodplains

Floodplains are defined as the land adjacent to a river channel, which is flooded periodically during high flows (e.g., Bridges 2009; Leopold and Wolman 1970; Rice 1949). Floodplains are geomorphological features built by deposition of fluvially transported sediment. Deposition of alluvial sediment occurs through two main processes, either vertical aggradation through deposition from overbank flows or lateral aggradation due to deposition on point bars.

Rates of overbank deposition are controlled by the return period of out-of-channel flows and by sediment concentrations in the flow. Return periods for out-of-bank flow are commonly in the range of 1–2 years (Wolman and Leopold 1957), although there may be significant variation around this typical figure (Knighton 1998) dependent on climatic regimes and floodplain characteristics discussed in greater detail below. Of more significance in controlling accumulation rates is variability in fluvial sediment load as a function of catchment character and human activity, which can lead to wide variations in rates of overbank deposition (Table 7.1). Vertical accretion of floodplains by overbank sedimentation leads to variation

Table 7.1 Rates of overbank deposition and overbank deposition as a proportion of catchment sediment loads for a range of global rivers.

River	Catchment Area	Sedimentation rate mm/yr	Floodplain sedimentation as % of SSL	Reference
Culm, SW England	276	Not reported	28%	Lambert and Walling 1987
Waal, Netherlands	Distributary of Rhine (Rhine = 220 000)	09–3.1	19%	Middelkoop and Asselman 1998
Severn, SW England	6850	Not reported	23%	Walling and Quine 1993
Coon Creek, Wisconsin	360	Not reported	50%	Trimble 1981
Lower Mississippi	3.1 m	Not reported	24%	Kesel et al. 1992
Amazon	6 m	Not reported	10%	Meade 1994
Amazon	2.2 m	Up to 3000	46–80%	Mertes 1994
Rhine distributaries	220 000	0.2–15	Not reported	Middelkoop 2002
Waipoa	1580	40–60	5%	Gomez et al. 2003
Tweed	4390	0–5.4*	39%	Owens et al. 1999
Ouse	3315	0–12.3*	39%	Walling et al. 1998
Wharfe	818	Not reported	49%	Walling et al. 1998
Strickland	75 000	10–55	Not reported	Alin et al. 2008
Upper Mississippi, various catchments	700–170 000	0.2–20	Not reported	Knox 2006
Galena River, Wisconsin	526	7.5–19	Not reported	Magilligan 1985
Yellow River	752 000	0–120	63%	Ran et al. 2014
Trinity River, Texas	43 000	18.5–45.4	53–98%	Phillips et al. 2004
Saskatchewan River	8000	1.5–3.9	Not reported	Morozova and Smith 2003
Danube River	Not reported	3–20	Not reported	Reckendorfer et al. 2013
River Elbe	148 000	0.2–15	Not reported	Krüger et al. 2006

*calculated based on assumed density of sediment of 1.3 g/cm^3

in rates of deposition across floodplain surfaces. Low points in the floodplain tend to be areas of preferential accretion unless they are flood channels. Over time, vertical aggradation of floodplains tends to increase the height of channel banks and so reduce the frequency of overbank inundation. However, floodplain aggradation is often accompanied by channel bed aggradation (e.g., Gomez et al. 1998) simultaneously increasing the connectivity between the channel bed and floodplain surface. Gomez et al. suggest that a dominance of vertical aggradation is a characteristic of river systems where significant increases in fine sediment load have occurred (e.g., due to changing land use or climate change).

Whereas growth of floodplains by vertical aggradation is an additive process, the production of floodplain sediment by lateral accretion involves a balance of sediment deposition on point bar surfaces against reworking of floodplain material into the fluvial system through erosion of meander bends. Lateral rates of meander migration measured by Hickin and Nanson (1984) and Nicoll and Hickin (2010) vary between 0.1 and 9.4 m/yr for a total of 49 rivers across the Canadian Prairies. Reviews of reported rates of lateral migration based on a global literature review by Hooke (1980) and Lawler (1993) provide estimated ranges of 0–792 m/yr, with the majority of studies showing rates of less than 10 m/yr and higher rates only occurring in the largest river systems. Data from the Amazon show lateral migration rates ranging between 1.3 and 35.9 m/yr (mean 8.9), with higher rates measured in reaches with higher sediment loads (Constantine et al. 2014).

Rates of erosion, and the balance between lateral and vertical aggradation, vary significantly between and across river systems. Leopold and Wolman (1970) suggest that for most systems, lateral aggradation is dominant, since continuous vertical aggradation would be self-limiting due to increases in channel capacity. However, under circumstances where sediment loads are increased significantly by upstream erosion, vertical aggradation can be dominant at short timescales (e.g., Trimble 1981).

The complexity of river channel behaviour in space and time (Schumm and Parker 1973) means that simple classification of floodplain form and behaviour is challenging. One of the more widely cited approaches was developed by Nanson and Croke (1992). Nanson and Croke suggest that 'the range of processes involved in floodplain formation is now so large, and the variety of floodplain types so diverse, that almost every case study might be seen to require a new model', p. 459. They proposed an approach to developing a classification of floodplain types, based around stream power and sediment type, which reflect the energy available for the stream to erode its banks and the degree of resistance to erosion imparted by the bank material. They classified floodplain systems into three principle categories which were: 1) high-energy systems (bankfull specific stream power > 300 W m^{-2}) with non-cohesive banks; 2) medium-energy systems (bankfull specific stream power 10–300 W m^{-2}) with non-cohesive banks; and 3) low-energy systems (bankfull specific stream power < 10 W m^{-2}) with cohesive banks. Under this broad classification, Nanson and Croke elaborate with 13 sub-categories of floodplain form, differentiated by a range of dominant channel and floodplain processes. Although it is not made explicit in the scheme, the selection of stream power and sediment calibre as key drivers of the floodplain classification mean that there is a geographical dimension to the classification associated with downstream fining of sediment loads, and systematic variation of specific stream power (Knighton 1999). The downstream variation in characteristic fluvial forms also implies that the persistence of floodplain landforms, and rates of turnover of alluvial sediment through the floodplain system, vary systematically in stream systems.

Residence Times

Carbon storage in floodplain systems is strongly controlled by the rates and nature of sediment accumulation as discussed above. However, as noted in the introduction to this chapter, to consider floodplains simply as carbon stores is to overlook a critical part of their function in the terrestrial carbon cycle. Stored sedimentary carbon is the residual of the floodplain carbon budget, but focussing only on this element ignores the role of floodplain vegetation in fixing carbon, and the potentially significant dissolved and gaseous carbon fluxes generated from the floodplain, that are associated with *in-situ* mineralisation of allochthonous and autochthonous organic carbon. These microbially controlled processes are typically rate-limited, so that net fluxes are related to the potential period over which carbon transformations occur. The persistence of floodplain sediments within the landscape, or the floodplain sediment 'residence time', defines the upper limit of the available time for this processing to occur, and so has the potential to significantly influence floodplain carbon balance.

Lateral aggradation of floodplain systems occurs in parallel with meander migration as point bar deposits accumulate. Rates of sediment accumulation therefore correlate with rates of channel change and bank height. Estimates of rates of change can be derived from rates of channel migration, measured either from historical mapping and imagery, or through dating of lateral deposits. Lateral aggradation on the inside of meander bends is associated with erosion of existing floodplain deposits on the outside of bends. Therefore, for laterally aggrading systems, a key metric in understanding the dynamics of the sedimentary system is the residence time of floodplain sediments, which is the time taken for complete reworking of sediments by meander migration at a given site. Very mobile channels are likely to have shorter overall residence times, even though rates of lateral aggradation may be higher.

An approximation of sediment residence time in floodplains being reworked by lateral migration can be derived from a simple comparison of the rate of migration with the available accommodation space in the form of the floodplain width. Hudson and Kesel (2000) compiled data from the reviews of meander migrations studies by Hooke (1980) and Lawler (1993) in order to derive a relation between drainage area and lateral erosion rate, as described in Equation 7.1:

Equation 7.1: $$E = 0.0575 \, DA^{0.342}$$

where E is lateral erosion rate (m/yr) and DA is drainage area (km²). They argue that this is a reflection of a general scaling of river catchments with area. Similarly, we might expect floodplain width to scale with catchment area as discharge increases. Notebaert and Piégay (2013) investigated this relation using digital terrain models of the Rhone River catchment in France. For the Rhone the derived relation was expressed as in Equation 7.2:

Equation 7.2: $$FW = 472 + 0.096 \, DA$$

where FW is floodplain width (m) and DA is drainage area (km²). Assuming that the residence time of sediment in the floodplain system can be estimated as FW/E (e.g., Aalto et al. 2008), a relation between drainage area and residence time can be derived as Equation 7.3:

Equation 7.3: $$R = 472 + 0.0965 \, DA \, / \, 0.0575 \, DA^{0.342}$$

where R is floodplain sediment residence time in years. Figure 7.3 compares these calculated values with a rather limited dataset of estimated residence times from the literature (Table 7.2). There is a reasonable fit to the theoretical trend with steep increases in residence time in the largest rivers but the data exhibits significant variation around the main trend.

The limited accuracy of the residence time scaling is perhaps unsurprising given that the catchment-area width relation is based on a single large river, and that as Notebaert and Piégay (2013) demonstrated, factors such as lithology, base level change, sediment load and discharge are also significant sources of variation in floodplain width. For example, the scaling of lateral migration rates derives logically from the relation between catchment area, stream discharge and stream power (although modified by slope), which means that this is consistent with the observation by Constantine et al. (2014) that meander migration scales with sediment load, but only for a constant sediment supply. However, the data do at a basic level support the trend of longer residence times in the largest rivers.

Another way to assess this scaling relationship is to compare the predicted pattern against empirical records of the preservation of floodplain sediments. Lewin and Macklin (2003) analysed a large dataset of radiocarbon dates within alluvial sequences, together with cartographically dated sequences. Their focus in this study was to consider the association of dominant channel processes such as channel migration, incision aggradation and channel stability, on the preservation of a sedimentary archive in fluvial systems. However, in the fluvial system, preservation of an archive is the equivalent of longer residence times, so that this study

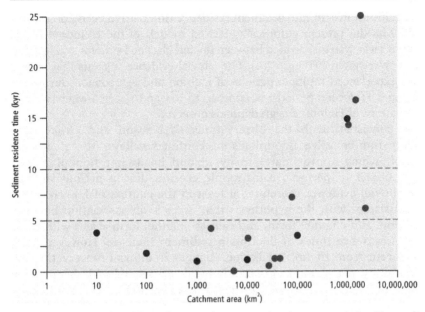

Figure 7.3   Relation of floodplain sediment residence time and catchment area calculated from scaling relations in Notebaert and Piégay 2013 (black circles) and Hudson and Kesel 2000, and comparison with data from Table 7.2 (grey circles). Source: Martin Evans.

Table 7.2   Reported floodplain sediment residence times from the literature. Note data marked * are from derived from Uranium series dating approaches in large catchments and may represent multiple phases of floodplain erosion in the catchment.

River	Catchment Area	Residence Time (kyr)	Reference
Strickland River, PNG	36 700	0.92	Aalto et al. 2008
Beni River	28 000	0.4	Wittmann and von Blanckenburg 2009
Amazon	2 200 000	7	Wittmann and von Blanckenburg 2009
Bella Coola, BC	5050	0.155	Desloges and Church 1987
Waipaoa River, NZ	2205	4.4	Phillips et al. 2007
Trinity River, Texas	46 100	0.67-1.13	Phillips et al. 2004
Upper Madeira River	1 050 000	$14 \pm 2^*$	Dosseto et al. 2006a
Andean Amazon	1540–21 197	$3 \pm 3^*$	Dosseto et al. 2006a
Mackenzie River	1 800 000	$25 \pm 8^*$	Vigier et al. 2001
Madeira River	1 420 000	$17 \pm 3^*$	Dosseto et al. 2006a
Andean Forelands	33 725–117 034	$7.6 \pm 0.7^*$	Dosseto et al. 2006b. 2008 2008 2006b

provides significant insight into sediment residence times at Holocene timescales. Lewin and Macklin present empirically derived models of the Holocene fluvial history of UK river systems, divided between upland glaciated systems and lowland unglaciated river systems (Figure 7.4). The alluvial evidence suggests that upland rivers have experienced multiple periods of incision and aggradation during the late-glacial and Holocene periods, in contrast to lowland systems where a more consistent pattern of Holocene aggradation is observed.

This is consistent with the observations of Nanson and Croke, that high energy non-cohesive floodplains in confined valleys (the floodplain classification which approximates many upland headwater floodplain systems) are subject to periodic catastrophic erosion during periods of high flow. The alluvial evidence therefore, at least in the context of UK river systems, is consistent with the hypothesis that there is an association between residence time and channel form, and that as channel form scales with basin area, mean residence times of floodplain sediment increase. However, what is also apparent from Figure 7.4 is that changes in alluvial preservation are also impacted by changes in sedimentation style associated with deglaciation (paraglacial sedimentation *sensu* Church and Ryder 1972). The implication is that the spatial pattern of characteristic sediment residence times is controlled not only by the discharge related scaling of basin size, but also by the influence of significant changes in sediment supply

Floodplain Geomorphology and Carbon Cycling

Carbon Cycling in River Floodplains

Floodplain residence times define the timeframe within which ecological processes drive floodplain carbon turnover. Figure 7.5 describes the carbon balance of floodplain systems. Three principal modes of carbon exchange can be identified: 1) inputs and losses from the floodplain system in dissolved and particulate form, associated with floodplain sedimentary dynamics; 2) inputs of carbon to the system through primary productivity; and 3) gaseous carbon losses due to soil respiration comprising root respiration and respiration by heterotrophic microbes.

Accumulation of allochthonous organic matter on floodplain surfaces is controlled by geomorphological processes, as described in the previous section. However, total organic matter inputs to the floodplain soils also include the litter from floodplain vegetation. Since floodplains are typically wet, nutrient rich sites, primary productivity in floodplain ecotones can be relatively high (Davies et al. 2008; Opperman et al. 2010). A full accounting of the floodplain carbon balance requires understanding both of the relative importance of allochthonous and autochthonous inputs of organic matter to the floodplain system, and of rates of OM turnover through heterotrophic respiration. These elements have been well

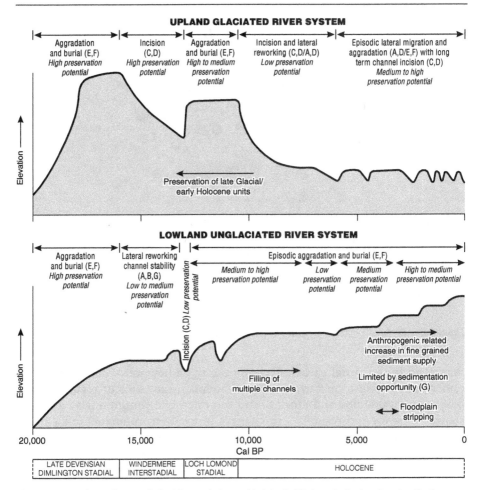

Figure 7.4   Holocene history of alluvial aggradation and incision in UK rivers, indicating the role of both catchment size (upland vs. lowland) and glacial history as controls on floodplain residence time. Source: After Lewin and Macklin 2003. Reproduced with permission of Wiley.

studied, but typically poorly integrated with geomorphological understanding of floodplain systems.

Controls on Microbial Carbon Cycling in Floodplain Soils

Rates of heterotrophic respiration of organic matter in floodplain soils are a key control on carbon accumulation. Where the NEE of these systems is shown to be negative (with respect to the atmosphere – a carbon source), this implies a further allochthonous carbon source. In this circumstance, the major role of the

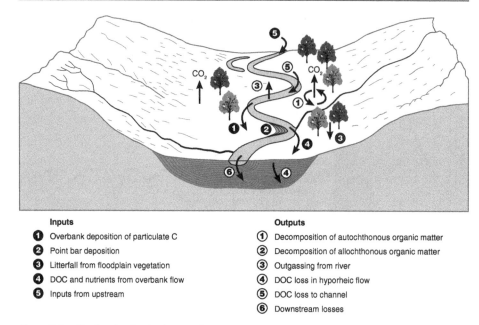

Inputs

1. Overbank deposition of particulate C
2. Point bar deposition
3. Litterfall from floodplain vegetation
4. DOC and nutrients from overbank flow
5. Inputs from upstream

Outputs

1. Decomposition of autochthonous organic matter
2. Decomposition of allochthonous organic matter
3. Outgassing from river
4. DOC loss in hyporheic flow
5. DOC loss to channel
6. Downstream losses

Figure 7.5   The fluvial carbon landsystem. Source: Martin Evans.

floodplain in terrestrial carbon cycling becomes a hotspot of carbon turnover rather than a long-term carbon store. Floodplains are sites of regular disturbance due to flooding and erosion, and soil microbial communities have been shown to have rapid responses to changes in local environments (e.g., Burns and Ryder 2001; Kobayashi et al. 2009), although microbial responses to disturbance in terms of changes in diversity and function are less well understood (Andersen et al. 2013).

The impact of flooding on microbial activity has been well studied in the context of a range of physical changes associated with inundation of floodplain soils. Perhaps the most fundamental change associated with inundation is the development of anaerobic conditions. Tate (1979) showed 65% reductions in aerobic metabolism in flooded soils. Initial increases in aerobic bacterial populations in the first 10 days of flooding were followed by declines and an associated increase in anaerobic microbial populations. Langer and Rinklebe (2009) similarly showed that inundation of floodplain soils on the Elbe river (Germany) led to reduced fungal activity. O'Connell et al. (2000), studying controls on decomposition of floodplain litter, showed that whilst the balance of microbial communities shifted during inundation (with fungi favoured in aerobic conditions and bacteria in flooded conditions), rates of litter decomposition were independent of oxygenation. They argue that whatever the flood conditions, a microbial community will develop to utilise the available carbon and promote litter decomposition. Song et al. (2008) found that bacteria were the dominant component of soil microbiota from a floodplain system. In this study, only 12% of the total microbiota was mycorrhizal fungi in aerobic conditions. Both bacteria and fungi were active decomposers in the presence of oxygen, but under anaerobic conditions bacteria were proportionally more important. The microbial soil community was found to be less abundant in anoxic conditions and diversity, although reduced, was less impacted. Changes in microbial community in response to flooding are in part dependent on the timescale of inundation. For example, in mesocosm experiments, Unger et al. (2009) showed that simulating 5 weeks of floodplain inundation did not significantly shift microbial community structure.

Microbial decomposition rates are limited not simply by oxygen status, but also by the presence of suitable substrate. For example, Craft et al. (2002) showed that sites distal to the river channel showed increased microbial respiration during flood events assumed to be due to an influx of suitable carbon substrate, whereas proximal sites were unaffected because substrate availability was not a limit on soil respiration rates in these better-connected locations. Substrate limitations are likely to be more significant in dryland systems, and McIntyre et al. (2009) have argued that flood events are temporal hotspots for carbon turnover in these systems, since between flood events both water and the availability of suitable carbon substrates are likely to be limiting.

Further evidence of substrate control on carbon mineralisation rates comes from the work of Rinklebe and Langer (2006), who showed that the ratio of microbial carbon to OM-derived carbon in floodplain soils is low in comparison to terrestrial soils. This implies that decay of plant residue is slower on floodplains, favouring the accumulation of organic matter. They argue that microbial communities are less efficient at using organic carbon on floodplains, which implies that this material is on average less labile than in terrestrial soils (consistent with the importance of deposition of more recalcitrant POC delivered upstream). This study also demonstrated that periodically inundated sites had higher microbial biomass than more seasonally inundated sites, in addition to a lower respiration rate per microbial biomass, which implied greater efficiency of microbial carbon use at drier sites. Basal respiration rates were however lower at the dry sites. This is in contrast to findings at other sites, which have linked influx of flood waters to peaks in respiration (due to OM decomposition) (e.g., Wilson et al. 2011), or have identified higher rates of microbial activity at drier floodplain sites (Kang and Stanley 2005).

The sedimentology of floodplain systems can impact substrate availability over different time frames. Rapid burial of surface soil layers can limit post-depositional diagenesis of alluvial deposits so that buried organic horizons in organic soils are sources of labile carbon (Gurwick et al. 2008). At longer time-scales, this process limits substrate availability and may result in the separation of autochthonous carbon (generated from primary production) from surficial microbial communities and ultimately drive carbon sequestration. In contrast, Wilson et al. (2011) developed a conceptual model of the impact of short-term floodplain inundation on microbial communities, organic matter decomposition and carbon mineralisation (Figure 7.6). This describes a step change in carbon mineralisation associated with flooding, caused by growth of opportunistic bacterial species and a diversification of substrate use, due to the inputs of fresh material to the floodplain system. Rates of mineralisation continue to increase for circa 14 days, with evidence of significant changes in microbial community structure 7 days after flooding. By 24 days post-flood, the microbial community is significantly modified from the pre-flood situation. These contrasting examples suggest that the impact of flood deposits on rates of carbon mineralisation is sensitively dependent on the rate and type of deposition.

Most studies of the role of microbial carbon processing in floodplain systems focus on the impact of inundation on microbial processes, through changes in the physical environment or changes in microbial substrate. One area that has received very little study is the role of the fluvial system in translocating microbes between environments. Wainright et al. (1992), studying the Ogeechee river (a blackwater river in Georgia, USA), demonstrated significant bacterial flux from floodplain soils to the river system, which supports high bacterial concentrations in the river channel. This is consistent with the role of floodplain soils as important sites of microbial turnover of organic matter. Little is known, however, of the

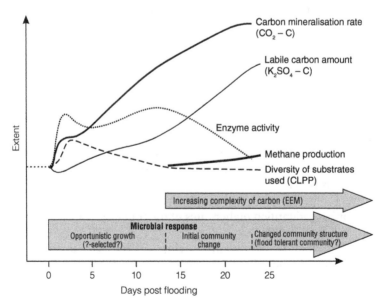

Figure 7.6 Microbial responses to floodplain inundation. Source: After Wilson et al. 2011. Reproduced with permission of Wiley.

impact of riverine microbial communities on the microbial functioning of flood-plain systems after inundation.

Overall, what emerges from the literature on microbial activity and decomposition of organic matter in floodplain soils, is that floodplain inundation favours bacterial over fungal heterotrophic communities and tends to lead to lower rates of carbon mineralisation. The nature and in particular the relative lability/recalcitrance of floodplain organic matter is also an important control on rates of mineralisation, with the relative importance of floodplain primary production (autochthonous OM) and fluvial delivery of POM derived from upstream sources (allochthonous OM) an important distinction in this context (Figure 7.7).

However, as the discussion above has illustrated, the interaction of multiple controls on heterotrophic microbial communities on floodplains and consequent patterns of carbon mineralisation, are complex in time and space. The ways in which microbial communities respond to physical and chemical changes in floodplain environments and the impact of these community changes on floodplain carbon cycling is an area which requires further research. This is an area that will require interdisciplinary collaboration between geomorphologists and microbiologists, and progress here will be vital to developing a process-based understanding of spatial and temporal change in the functioning of floodplain systems in the terrestrial carbon cycle.

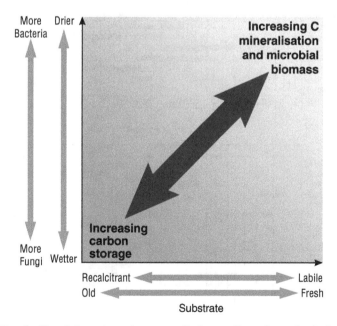

Figure 7.7   The role of inundation and organic matter quality in controlling carbon cycling by floodplain microbial communities. Source: Martin Evans.

Carbon Storage in River Floodplains

Sutfin et al. (2015) provide an excellent recent review of controls on carbon storage in floodplain systems. They identify three organic carbon pools on floodplains, namely soil organic carbon (SOC), large woody debris and standing biomass, and estimate that global storage of soil organic carbon on floodplains is 12–80 Pg of carbon, or up to 8% of total soil organic carbon storage. At the scale of the individual river floodplain, their review reveals significant variation across all three carbon pools (Figure 7.8), with values ranging over three orders of magnitude (approximately 1–1000 MgC ha^{-1}). In some forested systems, large downed wood can

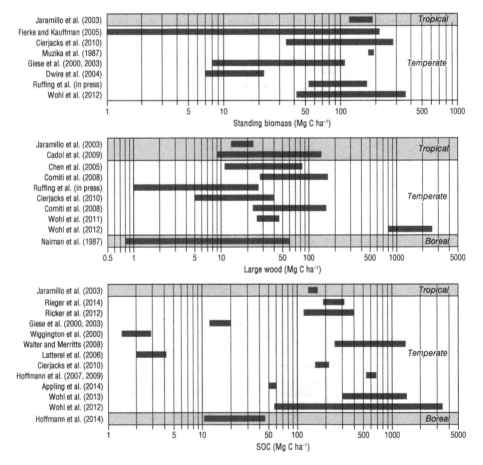

Figure 7.8   Reported magnitudes of organic carbon storage across three main floodplain carbon pools (Biomass, Large wood and SOC). Source: After Sutfin et al. 2015. Reproduced with permission of Wiley.

constitute the largest floodplain carbon pool (Wohl et al. 2012). Sutfin et al. (2015) also identify a range of macro-scale controls on carbon storage (Figure 7.9). These can be simply classified into controls on the magnitude of the carbon store, which are linked to rates and styles of organic sediment accumulation, the physical preservation of these stores (valley width, floodplain complexity and erosion) and controls on the preservation of carbon in the floodplain sediments (temperature and moisture), which in turn control rates of microbial turnover of organic matter, as discussed in the previous section. Sutfin et al. argue that this classification, derived from consideration of geographical variations in floodplain carbon storage, indicates that maximum organic carbon storage in floodplain sediments is associated with wet, cool environments, characterised by complex channel forms, unconfined valley systems and fine-grained sediments. A good empirical example of the role of floodplain geomorphology is work by Swinnen et al. (2020), who undertook detailed mapping of floodplain carbon storage in the River Dee in Scotland and demonstrated that floodplain slope and type were key controls with higher carbon storage in anastomosing and meandering reaches.

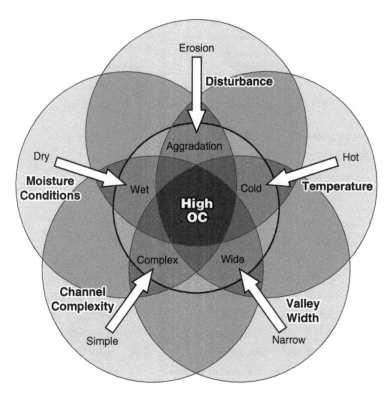

Figure 7.9   Classification of conditions favourable to organic carbon storage in floodplain systems. Source: After Sutfin et al. 2015. Reproduced with permission of Wiley.

The model presented by Sutfin (Figure 7.9) is consistent with the observation of Hoffmann et al. (2013b), that whilst carbon storage in headwater floodplains is limited by short residence times of sediment in dynamic channel systems, lowland floodplain systems are significant carbon sinks. Lowland systems are more typically unconfined (meaning there is significant accommodation space for carbon storage) and fine-grained (and thus are more cohesive). Table 7.3 describes documented rates of floodplain carbon storage from the literature with a range of carbon sequestration values from 0.7–665 gC m^{-2} a^{-1} with a median value of 110 gC m^{-2} a^{-1}. These values are consistent with the observation by Hoffmann et al. that floodplain systems in large rivers have carbon storage potential comparable to peatland systems. Their data on the Rhine river indicate that the highest rates of carbon storage are associated with increased sediment yields due to agricultural erosion. Total Holocene carbon storage in the lower Rhine basin is estimated at 1.1 PgC (Hoffmann et al. 2009), which is comparable to carbon emissions from land cover change over the same period.

Table 7.3   Carbon sequestration on floodplains based on data tables from Noe and Hupp 2005 and Sutfin et al. 2015, with additional data.

Study	Location	Floodplain carbon storage rates $gm^{-2}\, a^{-1}$
Brinson et al. 1980	N. Carolina, US	278
Noe and Hupp 2005	Atlantic Coastal Plain Rivers US	61–212
Hoffmann et al. 2009	Rhine, Europe	3.4–25.4
Alderson et al. 2019a	Woodlands Valley, Pennines, UK	110
C. Evans et al. 2013	Upper North Grain, Pennines UK	665
Craft and Casey 2000	Georgia, US	18
Zehetner et al. 2009	Danube, Vienna Austria.	100
Walling et al. 2006	6 Rivers in southern England	69.2–114.3
Ricker et al. 2012	New England	0.7–3
Tockner et al. 1999	Danube	290
Cabezas et al. 2009	Ebro River	140–300
Moreira Turcq et al. 2004	Amazon	100–250
Mitsch et al. 1977 and 1979	2 Rivers in Illinois	60
Bernal and Mitsch 2012	Old Woman Creek, Ohio	112
Bechtold and Naiman 2009	Modelled Washington State	36.4

A Floodplain Carbon Budget

Hoffmann et al. (2013b) identified a necessary synthesis of biogeochemical and geomorphological approaches to floodplain carbon cycling required to develop a quantified and process-based understanding of the controls on carbon flux. However, much of what has been written about the carbon balance of river systems has been focussed at the global scale (Aufdenkampe et al. 2011). Effective linkage of biogeochemical and geomorphological understanding will require a more process-based and nuanced understanding, which is likely to require integrated study, focussed at the level of the floodplain carbon balance. This approach demands data on the complete carbon budget of a range of floodplain systems. The floodplain carbon budget can be written as Equation 7.4:

Equation 7.4: $$\Delta C = PP + \left(D_o + D_b\right) - E - R - DOC - CH_4$$

where ΔC is change in floodplain carbon storage, PP is primary productivity, D_o is overbank deposition, D_b is deposition on point bars, E is bank erosion, R is soil respiration, DOC is flux of DOC from floodplain to channel and CH_4 is the flux of methane to the atmosphere from microbial methanogenesis. All units are gC m^{-2} a^{-1}. Unfortunately, given the logistical demands of measuring the complete budget, and the fact that most relevant studies have had more narrowly focussed aims, the range of available data on complete floodplain budgets is limited. However, whilst case study data are for the most part lacking, global data on the range of feasible values for key parameters are available. It is therefore instructive to construct a composite floodplain carbon balance, to assess probable ranges of values of the budget components, and hence the potential sensitivity of the total budget to variation in different elements. Table 7.4 presents such a composite budget.

The budget constructed in Table 7.4 yields a net carbon sequestration rate for the model floodplain of 943 gC m^{-2} a^{-1} (range 79 gC m^{-2} a^{-1} of carbon loss to 2960 gC m^{-2} a^{-1} of carbon sequestration). The largest fluxes are primary productivity and soil respiration, together with the terms relating to deposition and erosion of fluvial sediment. Losses of carbon from the system in dissolved form, or through methanogenesis, are relatively minor. The carbon balance in this composite budget represents a significant annual sequestration of carbon by floodplain sediment. In contrast, Hoffmann et al. (2013b) model typical floodplain carbon sequestration rates in the range 0–100 gC m^{-2} a^{-1} with best estimates for carbon storage in the lower course of the Rhine of 5.3–17.7 gC m^{-2} a^{-1}, values an order of magnitude lower than presented in Table 7.4.

In the budget presented above, the large sequestration rates are due to two budget components. Firstly, a large flux of carbon to the floodplain is associated with the overbank deposition term. This is based on the mean of values

Table 7.4 Based on a model channel with assumed floodplain width of 500 metres and a 2-metre depth of active bank erosion/deposition.

Carbon Budget Term	Relevant empirical data	Values	Reference	Assumptions	Carbon flux gC m^{-2} yr^{-1}
Primary Productivity	Direct PP estimates 1000 (g dry matter/m²/yr) excluding tropical Full range (500–3000)		Aselmann and Crutzen 1989	OC is 50% of organic dry matter.	−500 (−250 to −750)
Soil Respiration	Global range 60–1260 gC m^{-2} a^{-1}	Calculated from PP (SR = 1.24PP + 24.5)	Raich and Schlesinger 1992		645 (335–955)
C Deposition overbank	Mean overbank deposition 18.95 mm/yr (0–60)		Data from table × excluding Amazon data	Sediment density 1.3 g cm^{-3} and POC as 3.4% of SSC (0.2–10.1)	−838 (0 to −2652)
C deposition on point bars	Rate 4.75 m/yr (0.1–9.4)	Rate from 49 Canadian rivers and POC content from global dataset	Hickin and Nanson 1984; Ludwig et al. 1996b	POC is 3.4% of SSC	−420 (−8.8 to −831)
Bank Erosion	Assumed from rate of meander migration	Ditto	Hickin and Nanson 1984; Jobbágy and Jackson 2000	SOC is 1.1%	136 (2.9–269)
CH$_4$			Aselmann and Crutzen 1989	Value for tropical floodplains and so likely to be a maximum figure	19 (0–18.9)
DOC	Global fluvial DOC fluxes in the range 0–3 t km^{-2}	DOC fluxes calculated from dataset of global rivers	Ludwig et al. 1996b	Assuming approximately reduction by a factor of 10 between soil source and outlet (following Worrall et al. 2012)	15 (0–30)
Carbon balance			−943 gC m^{-2} a^{-1} (−2960 −79)		

from Table 7.1 (19.0 mm/yr), but these values are likely to be biased towards well-studied systems with high rates of deposition. Hoffmann (2013b) reports pre-human rates of overbank deposition on the Rhine of 0.5 mm/yr rising to 15 mm/yr in the period of human impact. Substituting these figures into Table 7.4 gives a range of sequestration rates between 127 and 768 gC m^{-2} a^{-1}. These values are still an order of magnitude above the measured long-term sequestration rates, implying that the majority of OC delivered to the floodplain by overbank deposition is mineralised (this indicates higher rates of soil respiration associated with greater substrate availability).

If significant mineralisation of deposited carbon is occurring on the floodplain, it has implications for the carbon balance associated with channel migration. Although a balance of erosion and deposition might be expected, associated with channel migration, the aged and decomposed material eroded from river banks is likely to be less carbon rich than the freshly deposited organic sediment load, so that higher rates of channel migration are also likely to lead to an enhanced floodplain carbon sink. This process is represented in Table 7.4 because of the different values for organic content of suspended sediment and floodplain soils which derive from the literature. If the model carbon balance is recalculated using minimum observed values of channel migration and overbank deposition, the floodplain system becomes a net carbon source. Therefore, the geomorphic instability of floodplain systems is an important driver of carbon storage. One potential flux not considered in this model is loss of plant litter from the floodplain as part of the particulate organic carbon flux. Inclusion of this flux into the model would reduce the magnitude of carbon storage. Litter can be a significant flux of POM to the river system, particularly during periods of autumn senescence of floodplain vegetation (e.g., Grubaugh and Anderson 1989), although measurements on Amazonian forest systems have suggested that the overall flux might be as low as 1% of litterfall (Selva et al. 2007).

The model presented in Table 7.4 is a first-order analysis of floodplain carbon balance based on very broad ranges of values for the key fluxes taken from the literature. As such, it cannot claim to be representative of any particular floodplain system. What it does demonstrate however, is that the lateral (erosion and deposition) and vertical (primary productivity and soil respiration) carbon fluxes from the system are typically of the same order of magnitude, so that changes in both the biological and physical components of the floodplain system have the potential to shift the balance between floodplains as sites of carbon storage and floodplains as sites of carbon turnover. The significant role of the balance of OC content of eroded and deposited sediment in this budget highlights the importance of the interaction of these two components. Understanding the full carbon budget of a range of river system types is an important research need in this area, which will require collection of new empirical data, which considers the dynamic components of the floodplain carbon balance.

Connectivity and Carbon Cycling in Floodplain Environments

Floodplains are ecotonal sites, which require insight from geomorphology, ecology and hydrology to understand temporal and spatial variations (Thoms 2003). McClain et al. (2003) argue that 'hot spots' and 'hot moments' of carbon cycling occur at the interface of terrestrial and aquatic ecosystems. The periodic inundation of floodplains distinguishes floodplain soils from hillslope soil systems, and the regular provision of fresh alluvium to floodplain soils, together with channel migration means that floodplain systems are characterised by significant heterogeneity (Bechtold and Naiman 2009).

The concept of connectivity is a useful approach to understanding the periodic shifts in the location of the terrestrial–aquatic boundary, which occur during floodplain inundation, and the associated fluxes of sediment and nutrients between the channel system and floodplain soils. Connectivity is a widely used concept in Geomorphology (e.g., Brierley et al. 2006) and across the biological and hydrological sciences (e.g., Wainwright et al. 2011), where analysis of contiguity and functional connectivity are central to an understanding of physical fluxes. Wainwright et al. (2011) term these two elements 'structural connectivity' and 'functional connectivity' respectively. Whilst the analysis of structural connectivity is well established in disciplines such as landscape ecology through approaches such as pattern analysis (Turner 1989; Uuemaa et al. 2013), an understanding of functional connectivity is a more significant challenge, requiring approaches that allow the tracing of materials through river systems (e.g., Koiter et al. 2013; Pfister et al. 2009) or the use of temporal modelling approaches (Wainwright and Parsons 2002). In river floodplain systems, connectivity can be defined in terms of the return period of overbank floods, so that the tools for analysis of connectivity exist, but the challenge of integrating understanding of sediment flux, biological activity and carbon cycling at the high temporal resolutions implied by flood events is an important research frontier.

Floodplain connectivity has been shown to influence carbon dynamics through its control on primary productivity, microbial communities and supply of allochthonous OM. Noe and Hupp (2005) used artificial marker horizons (feldspar clay) applied to floodplains on the Atlantic coast of the USA to study rates of carbon accumulation over periods of between 3 and 6 years, and demonstrated that reductions in connectivity between floodplain and channel lead to reduced accumulation of sediment and organic matter on floodplain surfaces. A similar finding emerged from work by Reckendorfer et al. (2013), who showed on the Danube that primary production was strongly linked to fluvial nutrient inputs in well-connected systems, but that autochthonous litter production was more important in floodplain areas that have become disconnected from the main channel.

Quantifying functional connectivity for floodplain systems involves consideration of the return period of overbank flows. As noted above, the typical return period for overbank flow is often taken as 1–2 years, but may be variable. The frequency of inundation at any given point on a floodplain may also be subject to fluctuations, and may change over time if deposition and channel migration change floodplain topography and proximity to the main channel. In a study of carbon accumulation in headwater systems in the UK, Alderson et al. (2019a) report changing patterns of flood deposits and carbon storage over time associated with vertical aggradation of a floodplain system. Zehetner et al. (2009) demonstrated high rates of carbon accumulation (up to 180 gC m^{-2} a^{-1}) associated with higher inundation frequencies (sub-annual) and decreasing carbon sequestration rates in older soils on the Danube floodplain, where floodplain connectivity has been reduced through flood control measures. They argue that these initially high rates of carbon sequestration are due to continual 'resetting' of soil formation processes, with delivery of flood sediments providing nutrients to support primary productivity and mineral material to promote organo-mineral complexation of OM. Burial of OM by overbank sedimentation also promotes preservation of floodplain carbon. The study suggests that microbial populations and OM cycling reach equilibrium in timescales of circa 100 years, where floodplains are disconnected from the main channel, with carbon sequestration rates decreasing at this point by an order of magnitude.

Modelling work of carbon cycling in forested floodplains in Washington State (Bechtold and Naiman 2009) supports these timescales and provides useful insight into the role of connectivity in controlling allochthonous and autochthonous organic matter production. Figure 7.10 shows model outputs and field data from the study, demonstrating that carbon storage increases rapidly in the first 100 years of floodplain development. In the model system, the floodplain is inundated for the first 40 years. Figure 7.10b shows the relative contribution of fluvially derived and autochthonous material. In this model system, fluvial deposition dominates only for the first 10 years of floodplain formation and accounts for 25% of total carbon accumulation at 25 years. Modelling approaches such as this, validated with field data, are a promising avenue for exploration of the complex interactions between inundation frequency, carbon turnover and sequestration, and the balance of allochthonous and autochthonous carbon production and storage in various river systems. Research to date supports the hypothesis that regular inundation promotes high rates of carbon accumulation, so that floodplain soils become sites of significant carbon storage. However, regular inundation also drives vertical accretion of floodplains and so provides a negative feedback to regular inundation. It seems likely therefore that carbon accumulation through this process is self-limiting (cf. Alderson et al. 2019a).

Connectivity in the sediment system is also important at much larger spatial scales. For example, at global scales, Leithold et al. (2016) contrast short steep rivers at active margins which export young carbon, and large rivers at passive

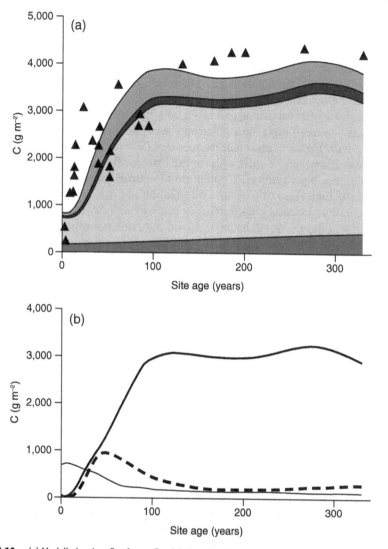

Figure 7.10   (a) Modelled carbon flux from a floodplain system (Dark grey = surface and root pool; black = active pool; light grey = slow pool; lines = passive pool; triangles = field data). (b) Relative contribution of autochthonous OM, fluvial OM and woody debris to carbon flux. Source: After Bechtold and Naiman 2009. Reproduced with permission of Springer Nature.

margins which export older carbon derived from erosion or metabolisation of long-term carbon stores in lowland river floodplains. The distinction is a function of sediment connectivity, and hence organic matter residence times are controlled by both lateral and longitudinal connectivity in the fluvial system.

The Geography of Floodplain Carbon Cycling – Towards a Model of Floodplain Function

The review of geomorphological and biological understanding of the cycling of carbon through floodplain systems presented in this chapter attempts to synthesise current knowledge of the complex interactions of biological and physical systems that control floodplain carbon cycling. Central to this understanding is analysis of temporal variability. However, a full understanding of the role of river systems in carbon cycling also requires a focus on spatial variation. Sutfin et al. (2015) have reviewed rates of carbon storage across the world's major climate regions and integrated their findings into their classification of the key controls on carbon storage illustrated in Figure 7.9. Another important axis of spatial change in river systems is scaling of fluvial processes with catchment area, highlighting functional differences across river systems from headwaters to estuaries.

Analysis of spatial variation at drainage basin scale is a standard approach in both biological and geomorphological work on fluvial systems, but to date this analytical approach has not been fully developed in consideration of floodplain carbon cycling. Classical geomorphological theory includes a focus on the morphology of drainage basin systems, with characteristic progression from headwater to lowland systems, and an association between downstream increases in discharge, and the development of a concave up long profile in fluvial systems (Gilbert 1877; Hack 1957). Many natural rivers conform to this pattern with rapidly declining slope from headwaters to the middle course of the river, and lower rates of gradient change as the channel approaches base level. Prior to 1950, the notion of river-long profile adjusting to discharge yielding curves of this form was commonly held (Knighton 1998). More recent work has emphasised the adjustment of channel form, slope and velocity to effect sediment transport through the river system, but retain the concept that discharge or basin area are important scaling factors (e.g., Figure 7.11a). Similarly, in the fluvial ecological literature following the work of Vannote et al. (1980) (influenced by the geomorphological work of Leopold et al. 1964), distance downstream (correlating with discharge and basin area) is considered an important control on the nature of nutrient cycling and food webs within the river system (Figure 7.12).

A significant complication with the notion of using discharge as a scaling factor in fluvial systems is the fact that discharge does not increase linearly downstream but is subject to step changes at confluences. Similarly, the material flux of the system can change significantly at confluences due to variation in the physical and ecological character of sub-catchments (Figure 7.11b). For example, Rice (1998) demonstrate the effect of tributaries on patterns of downstream fining in gravel bed rivers, and Rice et al. (2006) modelled the relation between discharge, sediment load and physical heterogeneity at confluences, to predict impacts on biological diversity. By the same token, Benda et al. (2004) hypothesised that

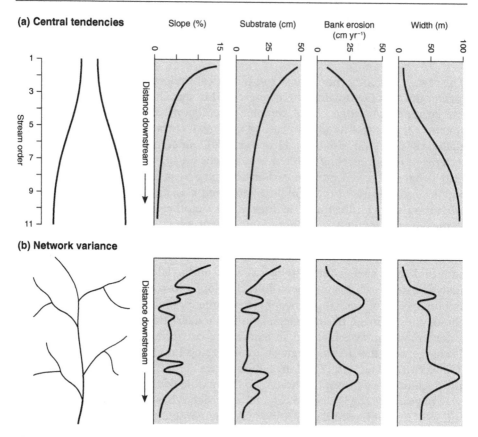

Figure 7.11   (a) Downstream change in the fluvial system according to classical theory. (b) More complex patterns of downstream change related to fluvial network structure. Source: After Benda et al. 2004. Reproduced with permission of Oxford University Press.

network structure is a fundamental control on diversity in channel and riparian habitats, and so a major control on biological diversity and productivity. Representation of monotonic downstream change in fluvial systems are therefore extreme simplifications, but represent a useful starting point for considering spatial change in river basins.

In the following section a conceptual approach to geographical patterns of floodplain carbon cycling is developed, building on these ideas, and on the discussion of floodplain systems in this chapter. Sutfin et al. (2015) focus their analysis of patterns of carbon storage around the potential for storage, and the physical and biological preservation of floodplain carbon. The analysis represents a spatialising of these ideas, combined with understanding of patterns of organic matter lability through the fluvial system.

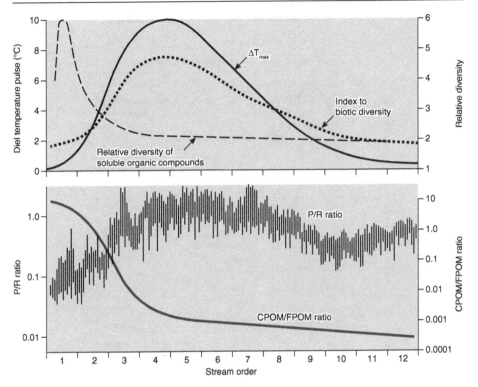

Figure 7.12   The River Continuum Concept. Downstream patterns of organic matter composition, photosynthesis:respiration ratio and biotic diversity. Source: After Vannote et al. 1980. © Canadian Science Publishing or its licensors.

A Conceptual Model of Carbon Cycling Through the Fluvial Sediment System

Figure 7.13 hypothesises downstream patterns of change in a number of key physical and biological parameters, which control patterns of floodplain carbon cycling. Of course, as noted above, systematic downstream patterns of change in parameters of the fluvial system are rarely replicated completely in real-world catchments. Changes in hydraulic parameters and sedimentology are strongly modified at confluences, caused partially by changes in bedrock configuration. Consequently, the conceptual model presented in Figure 7.13 describes broad trends in landscape-scale spatial patterns of carbon dynamics, rather than offering a description of any particular river system. Four key elements of the model: Physical setting, Physical preservation potential, Floodplain parent materials and Drivers of microbial decomposition, are discussed in more detail below.

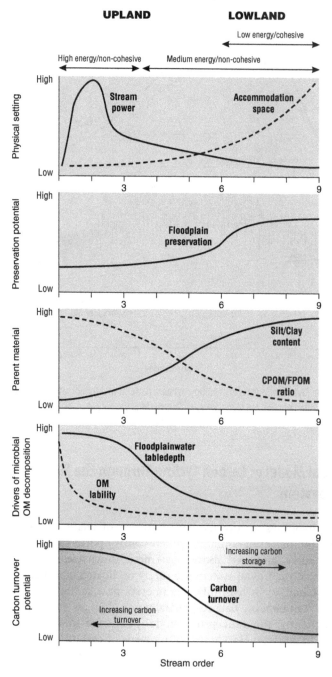

Figure 7.13   A conceptual model of carbon cycling through the fluvial system. Source: Martin Evans.

Controls on Variability in Floodplain Carbon Dynamics

Physical Setting

Specific stream power is assumed to peak close to the headwaters following Knighton (1999) and to decline downstream as channel slope decreases. In the classification of Nanson and Croke (1992), floodplains in low-order streams are therefore predominantly high-energy non-cohesive systems, where vertical aggradation dominates floodplain growth. Higher-order streams may be medium-energy systems with lateral aggradation, or low-energy cohesive systems particularly in distal distributary systems. Headwater valley systems are typically confined, so that there is limited accommodation space for the accumulation of floodplain sediments. In contrast, in unconfined systems downstream, the potential for accumulation of mineral and organic sediments increases non-linearly downstream. Consequently, floodplain volumes and the associated potential for carbon storage in floodplain sediments increases downstream.

Physical Preservation Potential

Long-term sequestration of carbon in floodplain systems is in part a function of the lifespan of the floodplain units. Nanson and Croke suggest that headwater high-energy systems are prone to catastrophic reworking. This is consistent with the observations of Lewin and Macklin (2003), which identify significant censoring of early Holocene alluvial sedimentary records ascribed to reworking of floodplain deposits. Lewin and Macklin suggest that in a UK context, upland floodplains have been subject to cut and fill processes and catastrophic reworking, whereas lowland floodplains have a longer history of aggradation. The potential for long-term sequestration of carbon in floodplains (e.g., Hoffmann et al. 2013a) is therefore preferentially a characteristic of large lowland systems.

Floodplain Parent Materials

Central to the River Continuum Concept of Vannote et al. (1980) is downstream decline in the coarse to fine particulate organic matter ratio in the river sediment load (CPOM:FPOM), controlled by biotic utilisation of the organic load. Of course, during flood events, pulses of particulate carbon are transported downstream and the available time for biotic decomposition of POM is limited so that the model is perhaps best conceived as a representation of average conditions. However, Pawson et al. (2012) demonstrated declining relative importance of POC loads during sediment transport events downstream of headwater peatlands. Downstream decline in POM concentrations are also driven by reductions in the relative contribution of steep headwaters to overall load as catchment size increases, so that a declining importance of CPOM is a reasonable assumption across a range of flow conditions. Similarly, downstream changes in mineral particulate load are usually characterised by fining, with higher proportions of silt and clay in the lower course of fluvial systems. Fine mineral material adsorbs DOC and the bound carbon is resistant to biotic degradation and oxidation (Kaiser et al. 1997). Silt and clay size fractions also form suspended sediment aggregates with fine organic matter, which promote deposition of fine material on floodplains (by increasing effective particle size and so increasing settling velocities) and physically protect organic matter from oxidation (Lal 2003; Nicholas and Walling 1996). Particle size controls on carbon stabilisation, therefore favour higher degrees of carbon preservation in higher-order stream systems.

Microbial Decomposition

Rapid reductions in OM lability downstream are predicted by the River Continuum Concept (Vannote et al. 1980) and are consistent with observations of DOM flux in UK river systems (Worrall et al. 2012). Loss of OM through rapid oxidation of the component delivered to the floodplain in high flow events is therefore likely to be maximised in headwater contexts, so that carbon turnover is favoured over storage in these systems. OM is preserved in anoxic conditions due to reductions in microbial decomposition. Consequently, sustained high water tables in floodplain systems favour carbon sequestration. Headwater systems typically have a flashier runoff regime and floodplain soils are typically coarse-grained in these high energy systems. More rapid drainage and a flashy hydrograph mean that headwater systems are less likely to sustain continually high water tables. Accordingly, headwater locations are likely to sustain conditions more favourable to rapid microbial decomposition of organic matter, with a progression to less favourable conditions downstream.

Carbon Turnover Potential in Floodplain Systems

Figure 7.13 suggests that the combined effects of changes in suspended sediment load character (organic and inorganic), floodplain preservation potential, floodplain sedimentary dynamics and the habitat for microbial decomposers all combine to indicate higher carbon turnover potential in headwater systems and greater propensity for carbon sequestration in lowland floodplains.

The pattern of OM lability described in Figure 7.13 assumes (as per the River Continuum Concept) that upstream sources of OM dominate floodplain carbon dynamics. However, significant *in situ* primary productivity on floodplains leads to a second input of relatively labile autochthonous OM. The balance of autochthonous and allochthonous carbon inputs to floodplains is partly controlled by frequency of out-of-bank flows and the connectivity of the floodplain area (Reckendorfer et al. 2013). However, the sedimentary dynamics of the system are also an important control. Floodplains which grow largely through lateral aggradation are likely to have a higher proportion of autochthonous carbon production in their sedimentary record, due to longer periods of *in-situ* organic soil development separating less frequent additions of sediment from overbank flows. Distinguishing allochthonous OM from autochthonous material in sedimentary records is a key research requirement for untangling the sedimentary carbon dynamics of floodplain systems.

Lowland floodplain systems which have the most dominant role in carbon sequestration are effectively wetland systems, where high water tables and high rates of primary production lead to *in-situ* storage of autochthonous carbon. In contrast, lower stream order floodplains may play a significant role in processing catchment carbon fluxes. Allochthonous organic carbon derived from upstream sources in these systems is rapidly cycled by floodplain biota and through reworking of the sedimentary system. Headwater floodplain systems have a high carbon turnover potential. They may therefore represent an important locus of mineralisation of fluvial carbon. In particular, where there are abundant sources of fluvial organic carbon in the catchment (e.g., forested or peatland systems), understanding the role of these types of low-order floodplains in processing carbon should be considered alongside carbon sequestration in lowland floodplains, to develop a complete understanding of fluvial carbon cycling.

Conclusions

This chapter has shown that floodplain systems are significant stores of terrestrial carbon. However, they are not passive receptors of organic matter. Floodplain systems actively cycle carbon under the control of both geomorphological and biological processes. The representative floodplain carbon budget calculated here suggests that a large proportion of the organic matter cycled through the floodplain system is mineralised, whilst the conceptual model discussed above suggests that there is a geography to this process with carbon turnover potential declining in a downstream direction.

Floodplain sediment stores are both sites of carbon storage but also reactors where carbon is turned over. Floodplain systems are potential hotspots of carbon transformation, so that understanding of the processes driving floodplain carbon cycling is important in the context of the global terrestrial carbon system. Geomorphological approaches to floodplain carbon cycling have tended to focus on carbon storage, and geomorphological insights into sediment budgets and the reworking of floodplain deposits within the fluvial system are an important contribution to understanding these systems. However, recognising the significance of carbon turnover at these sites requires a broader view of their functioning. Measured carbon storage is the residual term of large fluxes of carbon in and out of floodplain systems. As such, it is potentially sensitive to environmental change. A fuller understanding of the controls on floodplain carbon fluxes is required to assess the role of floodplain carbon storage in the terrestrial carbon cycle in a changing climate. This necessitates a research focus on integrating understanding of the physical processes that drive the formation and reworking of the floodplain system, in addition to the microbial processes which control turnover of carbon within the sediment store.

This chapter has provided a synthesis of current understanding and a framework for analysis of floodplain carbon systems. The scale of the floodplain carbon sink is significant in global terms, but more focussed empirical work integrating both the physical and biological dynamics of floodplain systems, and the associated fluxes of particulate, dissolved and gaseous carbon is required to develop approaches to understanding the fate of this sink.

Chapter Eight
Geomorphology and Carbon Cycling in the Coastal Ecotone

Introduction

The importance of rivers (Battin et al. 2009) and glaciers (Hood et al. 2015) as active conduits that process carbon from the terrestrial system and deliver it to the coastal zone is well established. The carbon flux to the coastal zone therefore represents an integration of carbon cycling further up the sediment cascade, as discussed in previous chapters. Particulate and dissolved carbon from the terrestrial system are important contributors to oceanic carbon pools, and the rate of carbon delivery has been significantly impacted by anthropogenic change in the terrestrial system. Regnier et al. (2013) argue that anthropogenic activity has increased the flux of carbon to inland waters by up to 1 Pg yr^{-1}, mostly as a consequence of soil erosion. The terrestrial organic carbon flux is an important carbon source for heterotrophic decomposers and can also contribute to ocean acidification (Bauer et al. 2013; Semiletov et al. 2016; Zigah et al. 2017). The near-coastal zone is the site of significant carbon transformation (Figure 8.1; Bauer et al. 2013). Many of these carbon transformations take place in the aquatic environment, so that the coastal zone can be seen as a dynamic site of microbial and photochemical transformation of organic carbon that has been delivered from the fluvial system. However, there is also an important role in carbon processing played by sedimentary environments at the land–sea interface, so that geomorphological understanding is integral to analysis of rates of carbon flux and transformation. As noted in Chapter 2, most of the mineralisation and burial of carbon, as well as

Geomorphology and the Carbon Cycle, First Edition. Martin Evans.

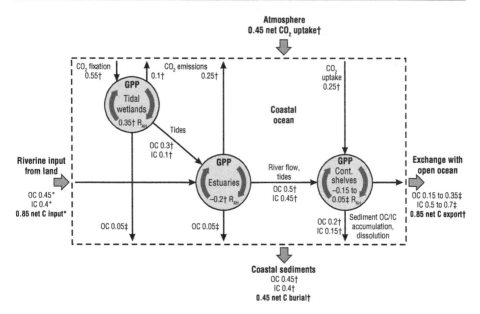

Figure 8.1   Carbon fluxes through the coastal system. Source: After Bauer et al. 2013. Reproduced with permission of Springer Nature.

carbonate deposition, occurs in the coastal zone at less than 200 m ocean depth (Bouillon et al. 2008). Oceanic carbon cycling is beyond the scope of this chapter, but this observation emphasises the critical role of the coastal ecotone in supplying terrestrial carbon to the ocean system.

Carbon exchanges in the coastal zone include fluxes of POC, DOC and DIC to and from coastal wetlands (e.g., saltmarsh and mangrove swamp) (Hyndes et al. 2014), transformation of terrestrial carbon in estuarine systems, and addition of terrestrial carbon to the coastal zone through coastal erosion. The outputs from this system are either sequestered (through carbon burial offshore), added to the oceanic organic carbon stock or mineralised and lost to the atmosphere. The elements of the coastal carbon system, and the impact of geomorphological processes as vectors of organic carbon across the coastal ecotone, are considered separately below.

Estuaries

Globally, rivers deliver around 0.9 PgC yr^{-1} (Ciais et al. 2013) to the coastal ocean. Most of this particulate and dissolved carbon is processed through the estuaries of large rivers. Estuarine systems are significant sources of CO_2 evasion to the atmosphere, with total losses estimated at 0.34 PgC yr^{-1} (Borges et al. 2005). Borges et al. note that the isotopic composition of aquatic shelf sediments suggests that they are less than 10% terrestrial in origin, so that the majority of fluvial sediment flux is mineralised in estuarine systems. Borges et al. further suggest that the approximate balance of fluvial carbon inputs and estuarine CO_2 flux to the atmosphere that they report does not indicate complete mineralisation of terrestrial carbon, because empirical evidence suggests that terrestrial organic carbon is both consumed and produced within the estuarine system.

Cai (2011) reviewed current evidence on estuarine carbon flux and showed that the high CO_2 flux from estuarine systems is disproportionate to their relatively small area, which is consistent with the observation that estuaries are turning over a significant carbon flux from large terrestrial catchment areas. Cai argues that the proportion of DOC decomposed in empirical studies of estuarine systems is insufficient to support the observed rates of CO_2 flux. Contrary to Borges et al. (2005), Cai argues that POC degradation rates cannot explain this gap, because residence times of estuarine water are too short to allow substantial microbial decomposition. Cai presents an alternative hypothesis to explain high CO_2 fluxes, whereby high DIC flux to estuarine waters is locally sourced from microbial decomposition of organic carbon in intertidal marsh habitats at the estuary edge. Bauer et al. (2013) suggest that both *in situ* primary productivity in estuaries and lateral export of carbon from estuary marginal wetlands can be of a similar magnitude to the riverine input of carbon to estuarine systems, so that carbon flux to the system is not simply a function of terrestrial catchment area.

Analysis of the carbon balance of over 60 estuarine systems on the east coast of the USA produced estimates that approximately 40% of organic carbon from both fluvial and wetland sources was either buried or respired in the estuarine system (Herrmann et al. 2015). Residence times, which are dependent on both the hydrology and geomorphology of the contributing catchment delivering runoff to the estuary, and on the geomorphology of the estuary itself, varied between 3 days and 39 months across the study area. This degree of variability has the potential to significantly impact both microbial consumption of organic carbon and burial of carbon in estuarine sediments (Abril et al. 2002). Across the whole dataset, the estuaries are net carbon sources to the atmosphere. This dataset also confirms the potential importance of estuarine intertidal wetlands, estimating that 35% of TOC comes from tidal wetlands, which constitute just 1% of the catchment area.

Such effective removal of organic matter from the water column in estuarine environments requires mechanisms for the rapid decomposition of organic molecules. As has been discussed for the fluvial system, two key mechanisms are photochemical and microbial degradation. Photochemical degradation occurs relatively rapidly, and total loss of DOC can exceed 50% (e.g., Moody et al. 2013). Commonly, loss of coloured DOC significantly exceeds this, with losses of 50–85% reported (Moran et al. 2000; Spencer et al. 2009), which implies that there is a proportion of uncoloured DOM that is resistant to photochemical breakdown. The nature of microbial control on decomposition of organic matter in estuarine systems is poorly understood.

Baker et al. (2015) analysed genomic sequences from shallow sediments from an estuary on the east coast of the USA, and showed that bacterial communities were complex with a dominance of uncultured lineages. These were most abundant in methane-rich zones of the sediment sequence, so that these bacteria were assumed to mediate carbon degradation and fermentation. Crump et al. (2004) analysed bacterial community composition in an estuary in Massachusetts, where estuary water residence time appeared to exert an important control on bacterial assemblages. The majority of bacterial types were either freshwater or saltwater species, transported into the estuarine environment by river flow or tidal action. However, where residence times fell below the doubling times of the bacterial groups (in summer and autumn), unique estuarine communities were identified. The influence of mixing of saline waters has also been observed in degradation rates, with Hernes and Benner (2003) showing photochemical degradation favoured at higher salinities and microbial decomposition at lower salinity values. Multiple studies have shown that microbial decomposition can be accelerated by physical alteration of the DOM through photochemical degradation (e.g., Moran et al. 1999, 2000).

As has been noted previously in this book, process-based understanding of the carbon cycle, here in the context of estuarine waters and sediments, requires better understanding of the interaction of microbial processes and abiotic controls on the aquatic and sedimentary system. Bauer et al. (2013) note that 'tremendous geomorphological differences between estuaries and a lack of synthetic modelling lead to considerable variability and uncertainty in estimates of estuarine carbon dynamics...' p. 64. Effectively, they postulate a direct link between estuarine geomorphology and carbon dynamics. The nature of this link however is poorly constrained. At its simplest, estuary morphology defines estuary hydrodynamics, which in combination with riverine inputs to the estuary, control the residence times of estuarine waters. Longer residence times favour greater microbial decomposition of organic matter and simultaneously deposition of particulate carbon. However, geomorphological control on estuarine processes goes beyond simple morphometry.

Hume et al. (2007) developed a four-level approach to classification of estuarine environments, based on abiotic components of the estuary and its catchment (Figure 8.2). At the top level, geographical variation in estuary function is distinguished based on latitude and location relative to large land masses. Below this, the hydrodynamic function of the estuary is classified based on morphometry and the interaction of river inputs and tidal variation. At the third level, the nature of the estuary catchment is considered, including land cover and geology. At the fourth level, local hydrodynamic processes are controlled by interactions of flows, with basin morphometry driving local deposition and erosion of sediment. Levels 2–4 of this classification represent the geomorphological differences bet-

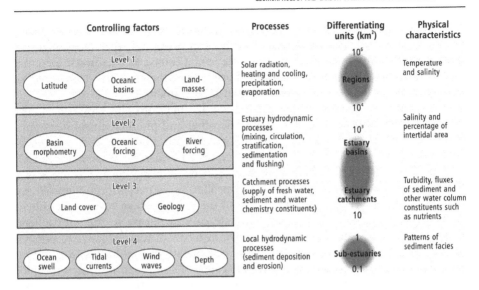

| Controlling factors | Processes | Differentiating units (km²) | Physical characteristics |

Figure 8.2   Estuary classification based on abiotic controlling factors. Source: After Hume et al. 2007. Reproduced with permission of Elsevier.

ween estuaries referenced by Bauer et al. (2013), including morphometry, the impact of upstream catchment processes on estuarine function, and erosion and deposition of sediments.

Hume et al. apply their classification to over 400 estuaries around the coast of New Zealand within a GIS framework. This approach has potential for upscaling local understanding of estuary carbon dynamics to produce better constrained estimates of estuarine carbon flux. However, this requires further work to link carbon turnover processes to the range of estuary types. What is clear is that there is potential to link geomorphological understanding of estuary form, and critically of the nature of the estuarine catchment, to a fuller understanding of these systems. In the Hume et al. classification, the representation of catchments is simple, being based on land use and geology, on the basis that these drive changes in sediment and water flux through the system.

An important avenue for further research in this area is investigation of catchment controls on the quality (lability) of terrestrial carbon delivered to estuarine systems, which will influence the potential of decomposition. The output of organic carbon at the mouth of a river system is an integration of the terrestrial carbon cycling processes operating in the catchment, as considered in the preceding chapters of this book. The potential impact of catchment geomorphology on the quality of organic matter delivered to estuaries has been examined by Moran et al. (1999), who analysed the lability of DOC feeding five estuarine systems in the USA and demonstrated considerable variability across the five systems. The most labile carbon and the lowest DOC concentrations were derived from steep Piedmont river catchments, whereas 'blackwater' river catchments with lower catchment slope and higher DOC concentrations were less labile. This is consistent with more rapid transport of fresh carbon through the steeper system. Similar results have been presented by Jaffé et al. (2004), who demonstrated that 'geomorphologically compartmentalised estuarine subregions', p. 195, could be distinguished based on spatial surveying of DOM quality.

To summarise, estuaries are predominantly heterotrophic systems, where microbial decomposition of POC, and microbial and photochemical decomposition of DOC, are sources of CO_2 to the atmosphere. Tidal wetlands are important sources of organic carbon and DIC export to estuarine systems drives CO_2 release. The relative importance of wetland and riverine carbon sources is a function of residence time and is thus controlled by estuary morphology. Where rapid flushing of riverine organic carbon to the coastal ocean occurs, the role of wetlands as a source of DIC may dominate carbon release to the atmosphere. Estuarine systems can appear conservative in terms of mass balance of organic matter, but this apparent stability masks complexity due to both consumption and production of carbon (Moran et al. 1999). The critical role identified for intertidal wetlands in the estuarine carbon cycle (Cai 2011) means that a full analysis of carbon cycling and geomorphology in the near coastal zone requires detailed consideration of these sedimentary environments.

Geomorphological Processes and Carbon Cycling in Coastal Wetlands

The role of vegetated coastal ecosystems (mangrove, salt marsh and seagrass beds) as zones of carbon exchange and carbon storage is significant. Globally, the area of these systems is one to two orders of magnitude below forests, but per unit area rates of carbon sequestration are considerably higher (Mcleod et al. 2011). Geomorphological processes play an important role in carbon storage in these systems. Trapping of suspended sediment and organic carbon during tidal inundation leads to deposition of organic-rich sediments. However, these 'blue carbon' ecosystems are being rapidly lost to anthropogenic interactions with the environment (Mcleod et al. 2011) and are threatened by changes in sea level predicted under conditions of global change. Three main ecosystem types dominate these coastal sites: mangrove swamps, salt marshes and seagrass beds. The role of each of these systems in coastal carbon cycling, and the relevance of geomorphological approaches to understanding carbon cycling in these contexts, is considered below.

Mangrove Ecosystems

Mangrove ecosystems occur in the intertidal zone throughout the tropics. The global area covered by mangrove forest (Figure 8.3a) is estimated at 138 000 km² (Giri et al. 2011), with the majority of these in a latitudinal band between 5°N

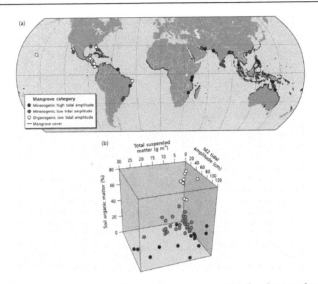

Figure 8.3   (a) Global distribution and classification of mangroves; and (b) Classification of mangrove types by tidal range and substrate character. Source: After Balke and Friess 2016. Reproduced with permission of Wiley.

and 5°S. They are dominated by a range of salt tolerant tree species, which are adapted to the low oxygen and high tidal range conditions of tropical coastal and estuarine environments. Mangroves typically grow on fine sediment accumulating in zones of reduced wave energy. The association between mangroves and fine muddy deposits in the coastal zone has meant that there has been extensive geomorphological study of mangrove systems.

Mangrove Geomorphology

In an important analysis of mangrove geomorphology, Woodroffe (1992) classified a range of key mangrove types including riverine, wave dominated, tide dominated, drowned coast mangrove and carbonate reef mangroves (Figure 8.4). The most widespread type are riverine mangroves, which are present in the intertidal zone of the estuaries of large tropical rivers. Most mangroves grow on muddy foreshore deposits, and are significantly influenced by flux of sediment from the fluvial system (riverine mangroves) or by two-way exchange of sediment with the coastal system (tide-dominated mangrove). In carbonate terrain where clastic sediment inputs are minimal the dominant sediment type can be peats developed from autochthonous primary production in the mangrove forest (Woodroffe 1992).

In mangrove systems, there is a close association between the height of the mangrove swamp, sediment accumulation and vegetation type, which may show vertical zonation across the tidal zone (Woodroffe 1992). The geomorphological classification of mangroves is based on topographic context in the coastal system.

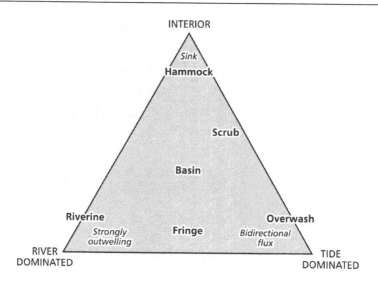

Figure 8.4 Classification of mangrove forest types in relation to physical processes. Source: After Woodroffe 1992. Reproduced with permission of Wiley.

Mangrove systems can also be classified functionally, but strong links between geomorphological process and ecological function in these environments means that there are clear relations between classifications based on function and those based on topographic context (Woodroffe 1992). For this reason, mangroves are commonly understood as an example of feedback between physical and biological processes controlling both the sedimentary environment and supporting vegetation growth at a site. In the classical biogeomorphological understanding of these systems, once physical conditions are suitable for mangrove colonisation, sedimentation rates are accelerated as the forest both reduces erosion through attenuation of wave and tidal energy, and promotes sedimentation through the trapping of particulate material in dense vegetation (Swales et al. 2015). Corenblit et al. (2015) argue that mangroves are good examples of 'engineer plants' that drive biogeomorphic feedback in contexts where vegetation interacts with flowing water.

Evidence for the role of vegetation in accelerating sedimentation comes from discontinuities in shoreline elevation profiles at the edge of the mangrove ecotone (Woodroffe 1992). However, an opposing view is that mangroves passively colonise suitable substrate, which is constructed through the physical sedimentological processes operating across the fluvial/estuarine system. Swales et al. (2015) aimed to test these ideas explicitly through high resolution dating of mangrove deposits in New Zealand. The study indicated that very rapid sediment accumulation typically precedes mangrove seedling recruitment, with limited evidence of accelerated sedimentation in vegetated areas. They therefore concluded that geomorphological processes driving estuarine sedimentation are the primary drivers of mangrove development.

Sediment accumulation rates in mangrove forests are relatively high, with reported values ranging from 0–40 mm yr^{-1}, with most values in the range of 1–10 mm yr^{-1} (Alongi et al. 2005; Swales et al. 2015; Woodroffe 1992). Adame et al. (2010) measured sedimentation rates in mangrove forest across a geomorphological gradient from riverine to tidal systems. No clear patterns of sedimentation emerged, although overall rates of deposition were highest in the fringes of tidal systems, which retained the majority of the tidal sediment flux. The proportion of terrigenous material retained in the mangrove sediments was highest in riverine mangroves.

Balke and Friess (2016) proposed a simple geomorphological classification of mangroves based on the tidal range at the site and the relative importance of mineral and organic sediments as a substrate for the ecosystem (Figure 8.3b). They argue that in systems with an organic substrate, biological controls on primary productivity are the principle control on sediment accumulation, whereas in mineral systems the allochthonous supply of sediment from river catchments or coastal sources is critical.

The biological and geomorphological dynamics of mangrove systems are an area of ongoing research, but what is clear is that these are systems where geomorphological processes and biological processes are closely coupled. A further aspect of this coupling is the role of microbial processes in cycling carbon from the stored organic sediments in mangrove systems. Carbon cycling and microbial processes are discussed further in the following sections.

Carbon Cycling in Mangrove Ecosystems

Carbon retention in mangrove systems is significant. Alongi (2014) estimates that mangroves contribute 10–15% of total global coastal carbon storage, despite representing only 0.5% of the area of the coastal ocean. They are also important sources of particulate carbon, exporting an estimated 10–11% of all POC to the ocean (Alongi 2014). Carbon burial is controlled by the rates of allochthonous organic sediment flux to the forest, the rates of autochthonous OM production through primary productivity and the rate of carbon mineralisation in the mangrove sediments. On a global basis, Alongi (2014) estimates that 58% of the soil carbon in mangroves is derived from autochthonous production, and 42% from allochthonous sources, including tidal advection and terrestrial carbon delivered by the fluvial system.

In Australian mangrove systems, Saintilan et al. (2013) estimate that recalcitrant carbon stored in mangrove sediments is predominantly derived from *in situ* accumulation, particularly from mangrove roots, whereas allochthonous organic matter deposited by tidal currents on the mangrove surface is more labile. In contrast, recent modelling of root decomposition rates for mangroves have indicated that just 24–29% of total carbon sequestration relates to root burial (Ouyang et al. 2017). Autochthonous production rates are high in Mangrove ecosystems. Bouillon et al. (2008) estimate that global mangrove primary productivity is $218 \pm 72\,TgC\ a^{-1}$, but that when comparing this to best estimates for carbon burial and mineralisation, half of this carbon fixation is unaccounted for. They suggest that this missing source is accounted for by high DIC losses in the form of dissolved CO_2 in streams draining from mangrove forest, a hypothesis which has been tested and not falsified by Maher et al. (2013) in Australian mangrove systems.

Productivity varies between mangrove types; for example, carbon budgets for riverine and coastal mangrove systems have shown that riverine mangrove production significantly exceeds production in coastal systems (Troxler 2013). It is clear from the literature that rates of carbon burial in mangrove systems are significant. Breithaupt et al. (2012) estimate that mangroves bury $163 + 40^{-31}$ $gC\ m^{-2}\ a^{-1}$, which represents 10–15% of mangrove productivity. Overall, carbon burial rates in mangroves have been estimated as circa $174 \pm 23\ gC\ m^{-2}\ a^{-1}$ by Alongi (2014). On an areal basis, this is a high rate of burial exceeding typical terrestrial peatland carbon sequestration rates by a factor of two. However, the impact of high carbon burial rates on greenhouse warming can be moderated to some degree by the methane and nitrogen dynamics of the systems. Chen et al. (2016) have demonstrated significant fluxes of CH_4 and N_2O from mangrove soils in Chinese systems, which reduce the greenhouse cooling impact of mangrove carbon sequestration by up to 33%.

Microbial Controls on Mangrove Carbon Cycling

High water tables in mangrove systems mean that mangrove sediments are largely anaerobic. Consequently, some studies have found an important role for autotrophic bacteria in carbon cycling. Sahoo and Dhal (2009), in a review of microbial activity in mangroves, present evidence for an important role for sulphate reducing pathways, which they argue account for almost 100% of mangrove CO_2 flux. Similarly, Alongi et al. (2005) reviewing mangrove carbon cycling, have suggested that mineralisation of carbon occurs at depth (up to 1 metre) in the mangrove sediments due to metabolism of root exudates from deep roots. Most of this oxidation is due to anaerobic sulphate-reducing bacteria, with limited contribution (5–12%) from aerobic respiration. In contrast, Kristensen et al. (2011), studying two Tanzanian mangrove systems, report that carbon flux from the surface of mangrove sediments is equivalent to depth integrated production in the upper 12 cm of the stratigraphy, with anaerobic carbon oxidation occurring in association with iron reduction, contributing less than 1% of total gaseous carbon flux. The nature of microbial control on carbon decomposition appears therefore to be spatially variable.

In addition to bacterial decomposition, Sahoo and Dhal (2009) identify an important role for fungi in initiating decomposition of mangrove litter and allowing secondary colonisation of these substrates by bacteria. Blum et al. (1988) estimate that fungal decomposition dominates mangrove detritus decomposition, accounting for around 80% of total losses. Although mangrove fungi have been widely studied, identification and classification is much further advanced than ecological understanding (Thatoi et al. 2013) and detailed understanding of the functional role of fungi in mangrove systems is limited (Hyde and Lee 1995). Mangroves have been identified generally as hotspots of fungal diversity (Thatoi et al. 2013). Hyde (1990) lists 120 fungal species with the higher fungi ascomycetes and deuteromycetes dominant in mangrove systems. The majority of mangrove fungal species are marine and saprophytic, driving initial decomposition of mangrove detritus through the release of extracellular degradative enzymes (Thatoi et al. 2013).

An active research area critical to fully understanding the processes of mangrove carbon cycling is understanding which carbon pools are decomposed to drive gaseous carbon release from mangroves. In addition to root exudates, algal biomass in surface sediments has also been identified as an important carbon substrate for autotrophic decomposers (Bouillon et al. 2004). Alongi et al. (2005) report high microbial biomass and rates of bacterial productivity observed in five Australian mangrove estuaries, which are interpreted as indicating that microbial biomass itself may be an important part of the carbon sink in some mangrove systems.

As with many of the sedimentary systems studied in this book, the understanding of the microbial processes which drive carbon cycling in depositional environments lags somewhat behind understanding of bulk rates of carbon flux and storage, and the processes which control movement of particulate carbon. The microbial ecology of mangroves is an active research area, and significant progress remains to be made linking understanding of population ecology to the functional role of these populations in carbon cycling.

Geomorphological Controls on Mangrove Carbon Cycling

Because of the major carbon fixing role of intact mangrove swamps, changes in the extent of mangrove systems due to erosion or deposition significantly impact carbon storage. Lagomasino et al. (2019) have demonstrated significant losses of carbon from African and Asian delta systems in the twenty-first century, which they ascribe to deforestation, but also to natural processes of erosion along mangrove creek systems. Woodroffe (1992) identified a large tidal range and strong currents as particular risk factors for the coastal erosion of mangroves, and these factors are exacerbated by sea level rise. Erosion of mangrove systems is therefore a potential positive feedback in the global climate system. Woodroffe (Figure 8.5) classified biological and geomorphological controls on mangrove growth and

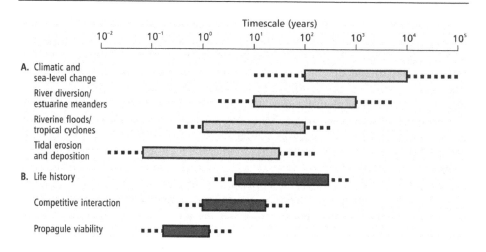

Figure 8.5   Characteristic timescales for processes controlling erosion and deposition in mangrove systems. Source: After Woodroffe 1992. Reproduced with permission of Wiley.

erosion. This classification suggests that change in mangrove systems, due to erosion and position, occurs at timescales spanning seven orders of magnitude, from days to hundreds of millennia. Human impacts can drive these processes both directly and indirectly.

Mangrove ecosystems are threatened environments, due to direct human intervention through deforestation. Alongi (2014) estimates that carbon losses of 90–970 TgC yr⁻¹ from deforestation now exceed global totals of carbon burial in mangrove systems. Indirect impacts include potential positive feedbacks from global climate change. Sea level change is a first-order control on deposition in mangrove systems and predicted levels of sea level rise will significantly impact mangrove dynamics. Organic mangrove systems can only avoid inundation where the vertical rate of mangrove peat accumulation exceeds rates of sea level rise, and so are dependent on levels of net primary productivity at mangrove sites. For minerogenic systems, accretion of new sediment to keep pace with rising sea level is a function of sediment delivery to the site, so that geomorphic feedbacks controlling terrigenous sediment production are critical (Balke and Friess 2016).

The persistence of mangroves over long timescales (Holocene and longer) implies resilience of mangrove ecosystems to sea level rise (Woodroffe et al. 2016). However, recent human impacts have the potential to push the system beyond conditions of local stability. Willemsen et al. (2016) identify two major drivers that have the potential to reduce the resilience of mangrove systems. The first of these is sediment starvation (cf. Balke and Friess 2016) and Woodroffe et al. (2016) particularly identify organic mangrove systems in areas of low sediment delivery such as carbonate terrains as being at risk. Similarly, Lovelock et al. (2015) suggest that under natural conditions, many mangrove systems can keep pace with sea level rise through sedimentation, but where catchment change (e.g., damming of river systems) reduces sediment flux, the mangroves can be rapidly overwhelmed. The second impact on mangrove resilience is related to coastal morphology and is termed 'coastal squeeze' by Willemsen et al. This is the situation where rising sea level and onshore topography mean there is no accommodation space for expansion of new mangroves, even if sedimentation rates are sufficient to support this.

Overall, in 70% of the global mangrove sites studied by Lovelock et al. (2015), sedimentation and peat growth were not keeping pace with rising sea levels, with the suggestion that areas with low tidal range and low influent sediment flux could lose mangrove ecosystems by 2070. Erosion of mangrove systems and loss of carbon sequestration potential therefore has the potential to be a positive feedback on climate change in the near term.

Salt Marsh Ecosystems

Salt marsh occurs in the upper section of the intertidal zone in temperate and high latitude regions (e.g., Vernberg 1993 and Figure 8.6). Salt marsh develops from the vegetation of coastal mud flats by dense stands of salt tolerant species, and consequently are favoured in locations where high fluvial sediment yields produce extensive deposition of fine sediment in sheltered shoreline locations such as estuaries and inlets (Figure 8.7). Salt marshes are typically dissected by networks of tidal creeks. In common with mangrove systems, the geomorphology

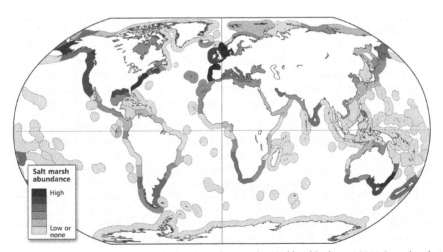

Figure 8.6   Map of global salt marsh abundance. Source: After Mudd and Fagherazzi 2016. Reproduced with permission of Cambridge University Press.

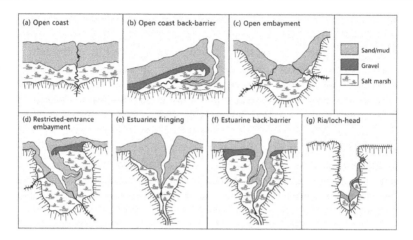

Figure 8.7   Morphological settings for coastal salt marsh. Source: After Allen 2000. Reproduced with permission of Elsevier.

of the salt marsh is controlled by interaction of vegetation cover, sediment supply and hydrodynamic impacts on the marsh system (Townend et al. 2011). Marsh vegetation promotes sedimentation through trapping of sediment and promotion of settling due to reduced flow velocities over rough marsh surfaces.

Salt marsh surfaces can aggrade rapidly in the early stages of their development as mineral sediment is added to the marsh surface. In conditions where falling sea level or reduced fluvial input limit minerogenic accumulation, salt marshes can become organogenic systems accumulating peat sediment (Allen 2000). Vegetation cover and reduced wave energy across salt marsh surfaces mean that vertical erosion of salt marsh is uncommon (Pethick 1992). However, lateral erosion of salt marsh systems is significant (Marani et al. 2011), driven by rising sea levels and human activity (such as wake erosion) (Theuerkauf et al. 2015). Ganju et al. (2017) studied 8 salt marsh sites in the USA and demonstrated that rapid lateral erosion meant that half the sites would be lost completely on timescales of 350 years. The key indicator of marsh expansion or erosion in this study was the ratio of vegetated to unvegetated marsh surface, providing an indication of the key role that marsh vegetation plays in the sediment budget of salt marsh systems.

Carbon Cycling in Salt Marsh Systems

Rates of carbon storage in salt marsh systems are high, with estimated global mean storage rates of 244 ± 26.1 gC m^{-2} a^{-1} (Ouyang and Lee 2014), comparable to rates recorded for mangrove systems. These estimates yield global carbon storage rates of around 10.2 TgC yr^{-1}. As is the case with mangrove systems, the high rate of storage is associated with relatively small total salt marsh areas, estimated at 41

637 km^2 (Ouyang and Lee 2014). Kelleway et al. (2016a) analysed the relative importance of geomorphic and vegetative factors in controlling carbon storage in Australian salt marshes. Stocks of carbon were found to be twice as high in fluvial settings as opposed to marine settings, but vegetation cover was not a significant control. Carbon storage was most closely linked to grain size, with carbon preservation maximised in fine sediment systems. Kelleway et al. interpret this pattern as indicative either of stabilisation of carbon in fine sediment systems or inferring high inputs of recalcitrant allochthonous carbon from fluvial systems. In contrast, Ouyang et al. (2017) have estimated 78% of saltmarsh carbon burial is associated with root burial (autochthonous production). As with mangrove systems, high rates of carbon accumulation are well established, but the relative importance of allochthonous and autochthonous carbon contribution is highly variable and less well understood.

Microbial Controls on Salt Marsh Carbon Cycling

The microbial populations of salt marsh systems have been studied in a range of contexts, but a comprehensive understanding of the nature of microbial control on carbon cycling remains a research objective. Anaerobic bacteria dominate near surface sediments and deeper sediments which are commonly anaerobic mostly support sulphur reducing anaerobes (as with mangrove systems). These groups have been demonstrated to play a significant role in organic matter decomposition in anaerobic sediments (Howes et al. 1984). Hu et al. (2014) studied bacterial processing of riparian and coastal wetland in the Lower Yangtze river system, and found reduced soil microbe activity in salt marsh areas compared to riparian wetlands. This was particularly connected with reductions in frequency of heterotrophic bacteria, indicating that lower soil respiration and increased carbon sequestration in these salt marsh systems is associated with a hostile environment for heterotrophic species. Analysis of lignin/cellulose decomposition has indicated that bacterial processes dominate over fungal processes (Benner et al. 1984), although complex interactions between fungal and bacterial processes have been postulated by Buchan et al. (2003), based on correlations between communities. Decomposition of lignin and cellulose by bacteria is an important source of dissolved carbon (Moran and Hodson 1989), but much of this is rapidly utilised by bacteria and only more recalcitrant components of this DOC contribute to production of humic compounds.

There is evidence that fungal groups also play a role in carbon decomposition in salt marshes. Dini-Andreote et al. (2016) demonstrated increasing stability of fungal diversity in late stage successional salt marsh, which they ascribed to a more constant supply of carbonaceous substrate from established vegetation. A number of studies have focussed on the response of microbial systems to pollution or human induced habitat change (e.g., Fernandes et al. 2017; Graves et al. 2016). However, in common with other environments, studies which link environmentally driven changes in microbial communities to observable change in salt marsh carbon dynamics are very limited. In a 2014 review, Darjany et al. (2014) state that '…indeed, salt marsh sediment bacteria remain largely in a black box in terms of their diversity and functional roles within salt marsh benthic food web pathways', p. 263.

Salt Marsh Systems and Sea Level Rise

Salt marsh systems are carbon storage hotspots (Ganju et al. 2017; Macreadie et al. 2013). Like mangroves, salt marsh systems are threatened by sea level rise and by human disturbance. Erosional losses of salt marsh habitat have the potential to significantly reduce net carbon storage, particularly where development of the coast behind the salt marsh limits salt marsh transgression in response to rising sea levels, so that the salt marsh fringe is narrowed. In these circumstances, beyond a threshold of marsh width erosion of the marsh can lead to the site becoming a net carbon source (Theuerkauf et al. 2015).

Geomorphological processes therefore become the key driver of carbon balance in the system. As in mangrove systems, rates of erosion in response to sea level change are significantly controlled by the ability of the marsh to aggrade, which in turn is linked to rates of sediment delivery to the marsh surface. Salt marsh systems where delivery of allochthonous material is limited, so that organic matter accumulation is dominant, are less able to keep pace with sea level change (Friedrichs and Perry 2001; Vernberg 1993) and are more prone to erosion or incursion of seawater.

The Salt Marsh–Mangrove Ecotone

Close similarities between mangrove and salt marsh systems are apparent from the discussion above, in terms of the geomorphic context of the sedimentary environments, biogeomorphological controls on sediment accumulation, high rates of carbon fixation and the sensitivity of intertidal sediment accumulation to changes in sea level. The two ecosystems occupy overlapping intertidal niches, and in some contexts, this association is apparent through ecotonal gradients between salt marsh and mangrove ecosystems (Kelleway et al. 2016b). The ecotone is typically temperature controlled due to the inability of mangrove species to tolerate subfreezing temperatures. In Florida, Doughty et al. (2016) have recorded expansion of the mangrove system into previously classified salt marsh areas due to climate change, and estimate a 22% increase in carbon storage associated with this change.

Seagrass Meadows and Carbon Storage

The third significant coastal ecosystem type that is significant in terms of carbon storage is seagrass beds. Seagrass species are macrophytes which grow and photosynthesise in shallow coastal waters where sunlight penetrates. They are highly productive ecosystems. Seagrass Net Primary Productivity (NPP) globally is estimated at 0.49 PgC, which is about half of total mangrove NPP globally (Mateo et al. 2006) and seagrass meadows are estimated to store 27.4 TgCyr despite occupying only 0.2% of ocean area (Fourqurean et al. 2012). Strictly, these are not terrestrial ecosystems, but they play a significant role in coastal carbon cycling and so are briefly considered here.

Seagrass meadows form submerged soils which are organic and largely anaerobic, and which may be significant stores of carbon. Soil formation is supported by the capacity of seagrass systems to trap detrital sediments from the shallow ocean, and the binding of these sediments by root systems produces cohesive soil systems which are able to resist erosion. Sediment trap data from seagrass meadows in Southeast Asia has indicated that circa 15% of trapped sediment (organic and mineral) is derived from the seagrasses, with the majority representing addition of sediment from the ocean to the seagrass meadow. Seagrass litter is decomposed by both aerobic and anaerobic bacteria, but the anaerobic conditions that exist in seagrass sediments mean that these are effective sites of carbon sequestration (Trevathan-Tackett et al. 2017). Fourqurean et al. (2012) estimate that global carbon storage in the upper metre of seagrass soils is 4.2–8.4 PgC, which is an order of magnitude greater than their estimate of standing carbon stock in seagrass macrophytes. Total seagrass soil carbon storage therefore exceeds both saltmarsh and mangrove environments.

Seagrass meadows are in decline globally (e.g., Short et al. 2014), threatened by a range of human impacts including eutrophication, mechanical damage, siltation and the impacts of climate change (storms and rising sea level) (Duarte 2002). A global review of studies of seagrass decline has estimated that 29% of known seagrass extent has been lost since 1879, with rates of decline of 7% per year since 1990 (Waycott et al. 2009). These losses of stored carbon to the ocean, if re-mineralised, could represent a total loss of 63–297 TgC/yr (Fourqurean et al. 2012).

In common with peatland systems, approaches to the restoration of seagrass beds have been explored with the potential to enhance carbon sequestration in damaged ecosystems. For example, Greiner et al. (2013) report large-scale restoration of over 1700 hectares in Virginia and demonstrated that carbon sequestration at rates comparable to intact systems could be achieved at timescales of around a decade (Gacia et al. 2003).

Marine ecosystem seagrass beds are not part of the terrestrial sediment cascade (although they may trap terrigenous sediments in the nearshore zone) and so are somewhat beyond the scope of this book. However, there are clear analogies with terrestrial organic soil systems and these systems represent a significant carbon sink within the coastal zone, which in common with mangrove and salt marsh systems, is threatened by human action. Enhanced erosion of all of these coastal organic carbon stores releases particulate carbon to the ocean system, which has the potential to be re-mineralised and lost to the atmosphere.

Coastal Erosion and Carbon Cycling in the Coastal Zone

The discussion of mangrove, saltmarsh and seagrass ecosystems above has emphasised the significant role of erosion in the carbon balance of these coastal ecotones. Erosion both limits the extent of these systems, reducing their net carbon sequestration potential, and also removes carbon rich sediments from long-term storage, releasing them to the aquatic system. Coastal erosion more generally also plays a role in carbon cycling in the coastal zone. Transfer of sediment and organic matter from the terrestrial system to the coastal ocean is normally understood as dominated by fluvial sediment flux from terrestrial catchments to the ocean via estuarine systems at the terrestrial-oceanic interface. However, direct transfers of material from the coastline to the ocean system occur where significant rates of coastal erosion are present. Where these eroded sediments are carbon rich, the net carbon fluxes can be large. These material fluxes are distinct from fluvial delivery of organic material to the coastal ocean, in that they are primarily controlled by coastal processes through the influence of tidal and wave energy impacting on erodible coastal sediments. On a global basis, this mechanism of organic matter transfer has not been widely studied. The best data come from analyses of rapid coastal erosion along Arctic permafrost coasts (e.g., Fritz et al. 2017). In these Arctic locations, high rates of coastal erosion coincide with the exposure of highly organic sediment at the coast, so that coastal erosion has increasingly been recognised as a key control on carbon flux to the ocean. At these sites, coastal erosion releases terrestrial carbon that has been stored for millennia in frozen peats and organic sediments.

Rates of coastal erosion along Arctic coasts have increased rapidly as a response to the melting of permafrost-rich cliff sections, which are in turn a consequence to climate warming of the Arctic. Reported rates of coastal retreat are significant. Along the Beaufort sea coast, rates of retreat in the range 20–40 m/yr have been reported (Obu et al. 2017). Across the whole Arctic Ocean, Fritz et al. (2017) estimate that the reduced strength of thawed coastal cliffs is generating average coastal retreat rates on the order of 25 metres per year in response to rapid warming of the Arctic. In these circumstances, coastal erosion can become the dominant input of particulate carbon to the coastal system. Along the eastern portion of the Siberian Arctic shelf, Semiletov et al. (2011) estimated that almost all of the carbon deposited derives from coastal erosion, with the inputs from the major Lena river system proportionally negligible.

For the whole Arctic Ocean, modelling based on carbon isotopic measurements by Vonk et al. (2012) indicates that the release of old carbon from the melting of coastal permafrost amounts to 44 TgC yr^{-1}. This is approximately three times the flux of particulate carbon to the ocean from the major northern rivers. They estimated that two-thirds of this carbon flux is lost to the atmosphere, with the remainder buried in near ocean sediments. The role of oceanic burial of old carbon from degrading permafrost regions has also been emphasised by Hilton (2017) in the context of fluvial sediment delivery from the Mackenzie River, where old POC flux (^{14}C dated to 5800 BP) is efficiently buried offshore creating a carbon sink.

Fritz et al. (2017) have summarised current understanding of the critical role of coastal erosion in contributing organic carbon to the Arctic Ocean under conditions of climate change (Figure 8.8). Fritz et al. note that coastal sediments influenced by permafrost constitute 34% of total global coastline. They estimate that the consequent flux of particulate organic matter to the ocean is $14\,TgC\,a^{-1}$, equivalent to the total annual flux of the major Arctic river systems. This estimate of coastal flux is lower than the results of Vonk et al. and the local data from eastern Siberia, but even this lower estimate establishes approximate equivalence between coastal erosion and fluvial sediment sources as an OM source to the Arctic Ocean.

Physical instability of carbon rich sediments typically causes major perturbation of local carbon budgets, with the loss of carbon stocks from sedimentary storage typically exceeding sequestration rates. However, as discussed in relation to peatland erosion in Chapter 5, the net effect on carbon loss to the atmosphere is strongly dependent on the fate of the particulate carbon. Rapid mineralisation of POC in the coastal zone generates a flux of carbon to the atmosphere, and a positive feedback to global heating. Where particulate carbon is efficiently buried in shelf sediments, the atmospheric impacts of the local geomorphological perturbation of the terrestrial carbon cycle may be neutral. The suggestion by Vonk et al. (2012) that two-thirds of the POC flux may be rapidly mineralised, implies that the erosion of coastal sediments in the Arctic is a positive feedback.

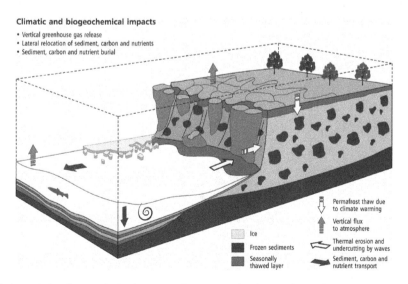

Figure 8.8   Carbon fluxes at the land ocean interface associated with permafrost melting on arctic ocean coasts. Source: After Fritz et al. 2017. Reproduced with permission of Springer Nature.

Of course, coastal erosion is not limited to the Arctic, but assessments of the impact on carbon cycling of the retreat of unconsolidated cliffs from other areas are extremely limited. However, some data are available from the rapidly eroding cliff lines of the east coast of the UK. In these locations, lightly consolidated glacial sediments in combination with rising sea levels lead to rates of coastal retreat, which approach those reported for the Arctic coasts. Typically, these glacial sediments are much less carbon-rich than Arctic permafrost deposits, but Lim et al. (2015) have highlighted rapid erosion rates in dune cliffs in northeast England, where large amounts of carbon are mobilised through the erosion of peat beds associated with the dune sequences. The impacts of erosion are two-fold, entailing both reduction in carbon storage through transfer of particulate carbon to the ocean, and reductions in carbon sequestration due to the removal of carbon accumulating ecosystems. Across the whole of the UK, Beaumont et al. (2014) estimate a trend of significant reductions in carbon sequestration due to observed and predicted coastal erosion rates for three coastal ecosystem types (Figure 8.9).

A comprehensive analysis of the role of coastal erosion in carbon cycling requires an integration of both of these mechanisms. The case of the Arctic Ocean and to a lesser degree the UK coast, together with the data on mangrove and salt marsh erosion reported above, indicate a need for a more compete synthesis of the role of geomorphological change at the coast in modifying carbon cycling at the land–ocean interface. Rates of coastal erosion are likely to accelerate in response to predicted rises in sea level, and the input of carbon-rich sediment to the ocean system, and the reductions in carbon sequestration by the ecosytems of the coastal fringe, are both potential positive feedbacks that require a fuller assessment at the global scale.

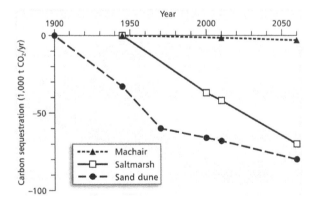

Figure 8.9 Carbon sequestration losses due to erosion of coastal ecosystems. Source: After Beaumont et al. 2014. Reproduced with permission of Wiley.

Carbon Fluxes at the Land–Sea Interface

Figure 8.10 is a representation of the carbon land system of the coastal zone, highlighting key carbon transfers and processes. The littoral zone is a dynamic sediment system, with fluvial and tidal influences on the sediment system overlain by the influence of long-term rises in sea level driven by global climate change. Two divergent elements of the coastal carbon cycle are apparent. Estuarine systems are hotspots of carbon transformation, acting as efficient reactors of terrestrially derived carbon delivered by the fluvial system. In contrast, coastal sediment systems, particularly coastal wetlands, are sites of carbon sequestration and storage, but also sites that are susceptible to erosion, which depletes both the carbon store and the sequestration potential. These contrasting roles are clearly illustrated in Figure 8.11 after Bauer et al. (2013), which summarises the role of elements of the coastal system in the global carbon balance.

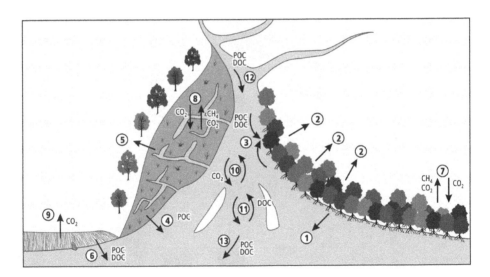

Figure 8.10   The coastal carbon landsystem. *Processes where geomorphological drivers dominate*: (1) Carbon loss by erosion/drowning of Mangroves; (2) Inland migration of Mangroves under rising sea level; (3) DOC and POC flux to Mangroves (Tidal and Fluvial); (4) Coastal erosion of saltmarsh sediments under rising sea level; (5) Landward expansion of salt marsh area due to rising sea level; (6) Input of POC and DOC to coastal ocean by coastal erosion. *Processes where Biological drivers dominate*: (1) Net carbon fixation by Mangrove ecosystems; (2) Net fixation of gaseous carbon by saltmarsh vegetation; (3) Mineralisation of carbon in coastal toe slope deposits; (4) Turnover of POC and DOC in estuarine waters; (5) Autochthonous production of DOC and POC in estuarine waters. *Integrated fluxes which are driven by geomorphological and biological processes*: (1) DOC and POC input from Fluvial system; (2) Flux of POC and DOC from estuary to coastal ocean. Source: Martin Evans.

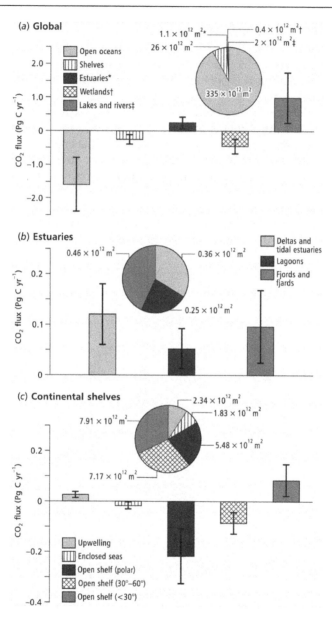

Figure 8.11   Contributions of coastal systems to global carbon cycling. Source: After Bauer et al. 2013. Reproduced with permission of Springer Nature.

A common thread of the chapters in this book examining carbon cycling through the sediment cascade has been the relatively poor integration of understanding of microbial processes (which determine rates of carbon transformation) with geomorphological understanding (which often determines the available timescale for those transformations). This pattern appears particularly acute in considering the coastal system. In some coastal systems, salt marshes in particular, the understanding of microbial ecology and of the role microbial processes play in biogeochemistry are well developed. Similarly, the geomorphological controls on coastal landforms and ecosystems are reasonably well understood. Where less progress has been made is in linking characteristic rates of microbial processing, with the periodicity of disturbance regimes in coastal systems. Current understanding emphasises geomorphological controls on carbon cycling, consistent with the dynamism of coastal sediment systems in a period of rising sea level.

After it has passed through the estuarine and littoral systems, terrestrial carbon is transferred to the coastal ocean with particulate carbon often deposited on the deltas of major river systems. As noted in the previous section, another recurring theme in this book is the fate of particulate organic carbon fluxes. Geomorphological processes erode, deposit and remobilise organic carbon, so that at times or in places of rapid geomorphological change, depletion of carbon stocks in one location can generate significant fluxes of particulate carbon and subsequent redeposition. If materials are redeposited in environments where microbial processes of decomposition and mineralisation are effective, then the original geomorphological change is likely to result in a loss of carbon to the atmosphere. In the coastal context, Blair and Aller (2012) have reviewed evidence on the fate of particulate carbon and identified three critical sedimentological contexts:

1) High-energy mobile deltaic muds in the aerobic zone, which lead to rapid mineralisation of particulate carbon.
2) Low-energy facies, where rapid accumulation leads to burial and preservation.
3) Areas fed by high relief terrestrial systems, where rapid rates of terrestrial erosion deliver high proportions of lithological organic carbon leading to significant preservation.

A consequence of these three characteristic carbon preservation facies, which is identified by Blair and Aller, is that the continental-scale geomorphological setting is strongly correlated with the efficacy of terrestrial carbon preservation in the coastal ocean. On active plate boundaries, high rates of erosion and high particulate fluxes lead to rapid burial and preservation of carbon, whereas low energy passive plate margins are hotspots for the aerobic mineralisation of organic carbon. Globally, Blair and Aller suggest that 27–38% of terrestrial organic carbon delivered to the ocean is buried.

The coastal system sits at the end of the terrestrial sediment cascade, and so integrates the range of geomorphological and biological processes which drive terrestrial carbon flux to the oceans. Previous chapters have shown the ways in which the geomorphological processes forming and reforming landforms and storing carbon, and the microbial processes which drive the cycling of that carbon, interact at a range of time and space scales, to determine net fluxes of carbon between the land, the atmosphere and the ocean. The continental scale geomorphological setting of the carbon stores at the end of these cascading systems plays a major role in the fate of terrestrially derived particulate carbon in the ocean system. This means that geomorphological control of the terrestrial carbon cycle occurs throughout the sediment cascade, and at the extremes of space scales, from hillslope facets to continental scale tectonic settings. Rising sea levels mean that the geomorphology of the contemporary coastal system is particularly dynamic, with the potential to release further stored carbon to the atmosphere in response to these changes.

Part III

A Geomorphological Approach to the Carbon Cycle

Chapter Nine
Geomorphology and Carbon Cycling in the Anthropocene

Introduction

Since the turn of the Millennium, the idea that we may be living in the 'Anthropocene' (Crutzen 2002; Crutzen and Stoermer 2000), an era during which human action is the predominant driving force in earth systems, has become widespread (Zalasiewicz et al. 2008). In fact, the proposal that the Anthropocene be defined as a new epoch in the geological timescale will be recommended to the International Commission on Stratigraphy in 2021 by the Anthropocene working group (Waters et al. 2016; Zalasiewicz et al. 2017).

The role of human impact on the form of the earth's surface is a longstanding area of study in geomorphology (Goudie 2013) and, in fact, Church (2010) has argued that geomorphology has evolved two distinct academic foci, one concerned with global scale earth system science, and the other concerned with applying geomorphological understanding to manage the human impact on the earth's surface. However, accepting the Anthropocene concept suggests a real overlap between these areas of study. The scale of human activity is such that understanding human interventions in the natural system has become a critical part of understanding the earth system as a whole (Figure 9.1). From a geomorphological perspective, Douglas and Lawson (2000) have estimated that global sediment flux associated with mineral extraction is around 57 Gt/yr, which exceeds fluvial sediment flux for the same period by a factor of three. Price et al. (2011) have argued that human action is the dominant geomorphological

Geomorphology and the Carbon Cycle, First Edition. Martin Evans.

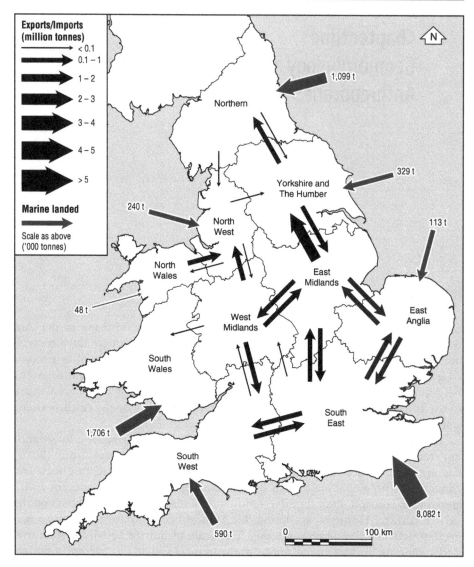

Figure 9.1   Flows of sand and gravel in the UK in 1997.   Source: After Douglas and Lawson 2000. Reproduced with permission of Wiley.

process operating at the earth's surface. More recently, Cooper et al. (2018) have estimated that total anthropogenic production of sediment has reached 316 Gt/yr, suggesting that anthropogenic production exceeded sediment production from natural sources circa 1955. Although various geomorphologists have queried the definition of the Anthropocene as a distinct geological boundary, largely because of the geographically time transgressive nature of many human impacts (Brown et al. 2013; Lewin and Macklin 2014), the overwhelming importance of human agency in contemporary geomorphological processes is widely accepted (Hoffmann et al. 2010; Syvitski and Kettner 2011; Walling 2006; Wohl 2013a).

This degree of human modification of the earth system implies that human agency has the potential to shift trajectories of ecosystem change at landscape scales. Over the past 20 years, the ecosystem services concept has become a dominant mode of understanding human interactions with the natural world (Costanza et al. 1997; Millennium Ecosystem Assessment 2005). The history of human interactions with the natural world is largely that of modification of natural systems for direct economic benefit, or of negative impacts on ecosystem health as a by-product of economic activity. The conscious acceptance of the scale and efficacy of human impacts implicit in the Anthropocene concept, in combination with an ecosystem service framework for understanding potentially positive ecosystem outcomes from human interventions, mean that the potential for landscape scale restoration projects aimed at maximising ecosystem services has been increasingly recognised (e.g., Aronson and Alexander 2013). In the context of terrestrial carbon cycling, the potential for active management of carbon in soils, sediments and biomass has been explicitly recognised in the Paris Agreement of the UN Framework Convention on Climate Change, which aims to develop common accounting approaches for Land Use, Land Use Change and Forestry (LULUCF) impacts on terrestrial carbon storage and flux.

Geomorphologists have been active in working with interdisciplinary teams on a range of landscape restoration projects including river restoration, peatland restoration and restoration of desertified land (e.g., Grabowski et al. 2019; Hevia et al. 2014; Shuttleworth et al. 2015). Not all restoration projects have carbon storage as a primary aim, but as significant interventions in the natural system, they often have carbon storage impacts. Effective management of terrestrial carbon storage requires understanding not just of the processes driving carbon flux, but also of the stability of the sedimentary systems that store organic carbon on land. This chapter explores the role of geomorphology in river and peatland restoration, in the context of impacts on terrestrial carbon cycling. It will also consider current understanding of carbon cycling in urban systems, which are hotspots of human impact. Whilst active interventions in the urban carbon cycle do not necessarily constitute restoration, they have the potential to be an important part of managing carbon in the Anthropocene and are considered here. The discussion will aim to identify the degree to which geomorphological understanding underpins the effective management of carbon cycling in these systems, which are sites where the geomorphological concerns with environmental management and with understanding the earth system intersect.

River Restoration and Carbon Cycling

Human impacts on global river systems are extensive and significant. They include flow regulation and dam building (Van Cappellen and Maavara 2016), pollution of river waters (Mekonnen and Hoekstra 2015; Meybeck and Helmer 1989) and modification of catchment land use impacting runoff and sediment production (Foley et al. 2005; Syvitski and Kettner 2011).

It has been argued that the scale of human impacts on fluvial systems is one key marker of the onset of the Anthropocene. Meybeck (2003) reasoned that river systems provide a critical connection between the atmosphere, pedosphere, biosphere and oceanic systems, and that the scale of human impacts has been such that they represent a significant modification of the Earth system (Figure 9.2). Effectively, river corridors have been transformed from natural- to human-dominated systems, and Meybeck identifies key modifications in river function including changes in water flux and pollution, but also changes in the carbon balance and greenhouse gas emissions.

An excellent review of the impacts of human activity on carbon cycling along river corridors has been presented by Wohl et al. (2017). They argue that two major impacts on river systems combine to increase fluvial carbon fluxes to the atmosphere and the ocean. Increased catchment soil erosion associated with agricultural use of floodplains enhances particulate carbon loss. Simultaneously, channelisation of river systems reduces the potential for carbon storage either in channel or on floodplains. However, reductions in longitudinal connectivity within channel systems due to impoundments mean that POC flux to the oceans is likely to be reduced in many river systems. Figure 9.3 from the work

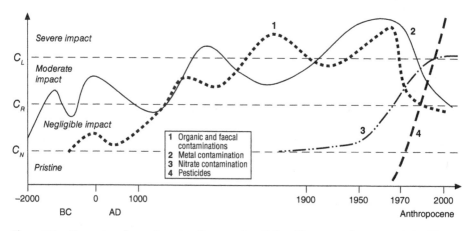

Figure 9.2   River system changes in western European rivers C_N, C_R and C_L are natural, recommended and limit parameter levels. Source: After Meybeck 2003. Reproduced with permission of The Royal Society.

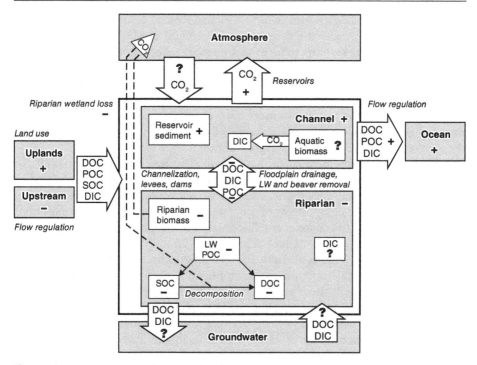

Figure 9.3   Human impacts on elements of fluvial carbon cycling. Source: After Wohl et al. 2017. Reproduced with permission of Wiley.

of Wohl et al. summarises their assessment of key processes driving the impact of human activity in fluvial systems. Overall, they argue that the net impact of human activity in river corridors is a reduction in carbon storage.

River restoration aims to improve the function of degraded river systems. The focus can be hydrological function, geomorphological processes or ecological integrity, or commonly components of all of these functions. Rivers have been manipulated for almost as long as people have lived on their banks, but restoration to more natural river function in modified systems is largely a feature of the last 40 years (Wohl et al. 2015). The scale of river restoration is significant; for example, Bernhardt et al. (2005) documented 37 000 river restoration projects across the USA, with annual expenditure of more than a billion dollars.

River restoration practitioners, and restoration science more generally, have debated appropriate targets for river restoration efforts. Often these focus on 'reference states' defined in relation to known historical river conditions (Nilsson et al. 2007; Tockner et al. 2003). However, Dufour and Piégay (2009) have argued that pristine reference conditions are both unknowable and unattainable, and argue for an approach to target setting that focusses on human benefits in an ecosystem service framework. Given the important role that fluvial systems play in both the cycling (Battin et al. 2009) and storage (Hoffmann et al. 2009) of terrestrial carbon, and the assessment from Wohl et al. (2017) that human modification of river systems reduces carbon storage, opens up the possibility of considering carbon sequestration as a targeted ecosystem service when restoring fluvial systems. Carbon sequestration has not been a primary driver for river restoration work, but significant work on the potential for carbon sequestration as one of the multiple benefits of more natural systems, represents an important nexus of geomorphological understanding of river systems and understanding of the terrestrial carbon cycle in an applied context.

Two modes of river restoration have been identified by Wohl et al. (2015): reconfiguration and reconnection. Reconfiguration alters the physical form of rivers to more natural conditions, whereas reconnection focusses on restoring connectivity between components of the fluvial system and involves management of flow regime as well as channel and floodplain geomorphology. Reconfiguration may not restore connectivity between channels and floodplains, and the hyporheic zone. Consequently, Wohl et al. argue that reconfiguration is less likely to succeed in restoring ecological function than reconnection. This also has implications for the degree to which these approaches can impact carbon storage and transformation. Reconfiguration may alter patterns of flow and sediment movement within the channel system, and therefore has some potential to impact rates of within-channel carbon turnover. However, relatively rapid transit times for water and sediment within channel mean that there is greater potential for an observed impact on carbon sequestration and turnover in the wider fluvial corridor.

As discussed in Chapter 7, floodplains play a major role in carbon cycling within fluvial systems, as sites where organic matter residence times are extended. Connectivity, controlled by river system morphology and flow regime, is an important control on these processes. Therefore, river restoration approaches which focus on reconnection of channels, either through hydrological interventions in the flow regime or geomorphological approaches to reconnecting the floodplain, are likely to promote natural carbon cycling and floodplain carbon storage. Channel–floodplain links can be important in stimulating autochthonous production on floodplains (Zehetner et al. 2009), as a result of the delivery of allochthonous organic matter to floodplain surfaces, where it may be buried or respired, and secondly, in flushing DOC from decomposing floodplain vegetation to the channel where it may support fluvial food webs (e.g., Tockner et al. 1999).

In a review of the attributes which create resilience in restored ecological systems, Timpane-Padgham et al. (2017) identify diversity and connectivity as important ecosystem characteristics that support system resilience to climate change, and also note that the ecosystem disturbance regimes play an important role in regulating natural systems. In the context of river restoration, Polvi et al. (2014) have classified the degree of geomorphic diversity in restored river systems based on sediment caliber, gradient, channel form and the presence of within-channel wood. They identify clear gradients of physical complexity between channelised, restored and pristine river systems in northern Sweden. There is an increasing recognition of the importance of understanding the interaction of biological and physical interactions in river systems, and the ways in which physical complexity underpins biogeochemical functioning (Wohl et al. 2015). Quantification of physical complexity can be based on spatial patterning of river features. It is this element of fluvial complexity that restoration by river reconfiguration primarily targets. However, a complete understanding of physical complexity has a temporal element, and includes the impact of flood disturbance on the fluvial system. The additional temporal element is incorporated by river restoration approaches that target reconnecting floodplain systems with the respective channels. In the following section, the way in which these approaches can impact on carbon cycling in the fluvial system are explored in more detail, in order to assess the degree to which river restoration can or should consider carbon fluxes and carbon sequestration as a feasible restoration target.

Carbon Management Through Channel Reconfiguration

River reconfiguration typically focusses on river channels, and may involve manipulation of channel form to achieve a more natural configuration, or to enhance the ecological status of the channel system. Channel modifications may include construction of meanders or addition of boulders to the channel to enhance habitat heterogeneity (Palmer et al. 2010). Such changes are likely to modify sediment storage in channels and may impact on carbon cycling. Lepori et al. (2005) demonstrated that restoration of channelised forest streams through the addition of boulders to the channel leads to greater retention of coarse particulate organic matter in channel beds, which supported fluvial food webs. However, even in restored systems, the magnitude of sediment storage and sediment residence times are likely to be significantly less than across the wider river corridor, which includes floodplain sediments and vegetation. However, river restoration that simultaneously restores more natural channels and floodplain systems has a more general possibility to enhance carbon storage potential. Brown and Sear (2008) argue that natural headwater streams (order 1–3) are stronger carbon sinks because natural multi-thread channel systems support floodplain wetlands with potential for long-term carbon storage. Increased sediment flux in European rivers due to agricultural change has led to a dominance of single thread channels and reduced storage. Brown and Sear suggest that whilst restoration in densely populated systems cannot go back to pre-impact conditions, a partial return to some anastomosing sections through restoration would enhance carbon storage.

An aspect of channel reconfiguration that may have a very significant impact on carbon cycling is longitudinal connectivity controlled by artificial impoundments. Because of the importance of dams in sediment retention and carbon storage (e.g., Butman et al. 2016), river restoration through dam removal (Magilligan et al. 2016) has the potential to have a major impact on carbon cycling. In the USA, over 60 dams a year are being removed, enhancing downstream connectivity within river systems. Dam removal can lead to substantially enhanced loads of dissolved and particulate organic matter downstream (Bednarek 2001), although Riggsbee et al. (2007) have argued that concentrations do not significantly exceed those observed in major floods on the same river system. Dam removal may also impact gaseous carbon exchanges by reducing methane flux from reservoir sediments (Battin et al. 2009). The net impact of dam removal on fluvial carbon cycling is potentially to drive increased carbon loss from sediments released into the oxic river environment; however, estimates of carbon loss to the atmosphere are highly dependent on understanding the eventual fate of reservoir sediments. These may be removed from site, stabilised *in situ*, or released to the river system through gradual or rapid dam breaching. Progress in this area will require understanding of both geomorphological (e.g., East et al. 2015; Randle et al. 2015) and biogeochemical (e.g., Gold et al. 2016) responses of the sediment system to restoration. This is an area where further work is required to understand impacts on carbon cycling.

Whereas in larger river systems dam removal can be viewed as river restoration, in higher-order streams river restoration may involve the addition of small dams through natural or artificial processes. Two approaches are of particular significance here. The first is the addition of large woody debris to stream systems. This is a widely used restoration approach (e.g., Larson et al. 2001; Shields et al. 2003), which aims to diversify stream morphology and habitat. A common assumption is that this will enhance biodiversity and that the increased complexity of the river system will enhance overall resilience of the restored system (Palmer et al. 2010). However, Palmer et al. note that physical diversity of river form has often not led to measurable changes in biodiversity. In a study of 78 river restoration projects involving addition of woody debris, only two showed statistically significant change in stream biodiversity. The impact of woody debris on carbon cycling is therefore likely to be more closely linked to carbon storage in the debris than to changes in stream biogeochemistry. Large woody debris (LWD) plays a significant role in within-channel carbon storage and carbon cycling. LWD decomposition can be an important source of POC and DOC to downstream systems (Chen et al. 2005; Seo et al. 2008) and also a significant carbon store, persisting for hundreds of years in river systems (Chen et al. 2005; Guyette et al. 2002, 2008).

The second connected river restoration approach, which may be significant in relation to small dams in river systems, is based on the critical role that beavers play in the geomorphology of many headwater stream systems (Burchsted and Daniels 2014; Gurnell 1998; Johnston 2014). Woody dams caused by beavers felling trees in headwater catchments have a significant impact on the geomorphology of headwater systems. Beaver dams drive heterogeneity of sediment storage in the channel system (Burchsted et al. 2010) with poorly understood implications for in-channel nutrient cycling. Frequent channel obstructions and the associated patchy sediment storage characteristics of beaver systems mean that traditional approaches to understanding network-scale fluvial biogeochemistry such as the River Continuum Concept are difficult to apply (Burchsted et al. 2010). The local impacts on carbon cycling can be significant. The pools behind beaver dams have been shown to be important areas of carbon turnover, with net carbon losses in excess of 200 gC m^{-2} a^{-1}. However, the potential impacts of beaver dams on fluvial carbon cycling extend beyond consideration of in-channel impacts. More significant impacts on catchment carbon balance result from the potential for carbon storage associated with beaver modification of sedimentation patterns in upland valleys.

Beaver dams are integral to the formation of alluvial valleys in headwaters, because the dams promote out-of-channel flows during storm events and consequently initiate sedimentation on floodplain areas (Westbrook et al. 2011). Channel avulsion in these circumstances can lead to relatively stable multi-channel systems, which extend the influence of the riparian zone across the entire width of the valley floor (Polvi and Wohl 2013). Given the potential of the riparian zone to act as a hotspot for carbon cycling (Chapter 5), this has implications for carbon cycling at the landscape scale.

Meadow soils associated with beaver dams (beaver meadows) have been shown to be areas of significant carbon storage, exceeding carbon concentrations in surrounding forest soils (Johnston 2014). Wohl (2013b) demonstrated that up to 23% of landscape carbon storage in Rocky Mountain headwaters was in beaver meadow environments, but that desiccation of sites where beaver populations had declined reduced this proportion to 8%. Beaver modification of the channel system and therefore of the geomorphology and sediment storage potential of headwater systems appears to play an important role in promoting storage of carbon in the fluvial system.

Understanding of the role of beavers in headwater systems has implications in the context of river restoration and carbon cycling. River restoration approaches may mimic the impact of beavers on the channel system. DeVries et al. (2012) describe a river restoration approach applied in a headwater stream system in Idaho, which involves construction of small log dams that promote out-of-channel flow during high flow events. The study demonstrates how these structures stimulate natural beaver activity in the landscape, leading to further restoration of connectivity between the channel system and the floodplain through out-of-bank flows at high discharges. Similarly, Pollock et al. (2014) propose that beaver conservation either by removal of external threats, or by direct habitat modification, is a preferred option for the restoration of river systems where channel incision has led to reduced connectivity between channel and floodplain, and the loss of fluvial wetlands. A combination of river engineering and habitat engineering has the potential to enhance carbon storage in the fluvial corridor.

Connectivity between the channel and floodplain system is a key control on the biogeochemistry of the floodplain as a landform. The role of the flux of nutrients and sediments to and from floodplains has been widely recognised through the Flood Pulse Concept (Junk et al. 1989), which emphasises the role of floodplains as local sources of organic matter to river systems, and as active sites of nutrient turnover. More generally, Burt and Pinay (2005) have emphasised the connection of hydrological flows and biogeochemical responses at the landscape scale, identifying the near channel zone as a 'hotspot' of biogeochemical change.

Carbon Management Through Engineering of Channel Connectivity

In the previous section, changes in floodplain connectivity in beaver impacted systems occur as a consequence of channel modification. In the lower courses of large river systems, connectivity has been dramatically reduced through the urban development of floodplain land and consequent embankment and other flood defences. In this context, river restoration approaches may focus on restoring connectivity through modifying these flood defences and incorporating floodplains into flood alleviation schemes.

Newcomer et al. (2012) studied organic carbon and nitrate dynamics on the floodplains of restored and urbanised stream systems in Maryland, USA, and demonstrated that autochthonous sources of OM dominated in urbanised systems. In better-connected restored sites, OM was dominated by more recalcitrant riverine organic matter. In this context, the connected system is more likely to store carbon, whilst the disconnected system turns over relatively labile autochthonous carbon. On the River Thur in Switzerland, Samaritani et al. (2011) characterised carbon pools and gaseous carbon fluxes in floodplains adjacent to both channelised and restored river reaches. A greater range and heterogeneity of carbon pools and fluxes was observed in restored reaches alongside increased biodiversity. Restored reaches included both gravel bars and more stable alluvial floodplain segments, and the greater physical and biochemical complexity of the system was interpreted as an indicator of functional diversity driven by greater connectivity and periodic inundation.

Reconnection of floodplains on very large river systems has been shown to generate similar results. Restoration work on the River Danube downstream of Vienna (Austria) has aimed to reconnect the floodplain and channel through the creation of low points in flood defences (which allow floodplain inundation at high flow) and construction of berms to maintain low water levels across some areas of floodplain between events (Welti et al. 2012). Analysis of floodplain respiration has demonstrated that potential microbial respiration is higher on the disconnected floodplains, with rates controlled by the nature of available organic matter and surface area, which is ultimately governed by floodplain morphology (Welti et al. 2012).

The close coupling of floodplain connectivity and carbon cycling identified in a range of fluvial contexts means that an understanding of the floodplain geomorphology at a range of scales, and of the role of channel connectivity in supporting autochthonous carbon production on floodplains is required in order to develop river restoration schemes which maximise carbon sequestration.

Understanding the relative importance of allochthonous and autochthonous organic matter in floodplain systems is key to the analysis of carbon cycling. A number of studies have demonstrated that river restoration approaches that modify floodplain connectivity can alter this critical balance. Changes in connectivity alter the frequency of floodplain inundation, so that chronosequence approaches to assessment of floodplain carbon dynamics can be a profitable avenue for research. Cabezas and Comín (2010) demonstrated that in agricultural areas of the Ebro river floodplain (Italy), sites with soils aged >60 years had significant autochthonous input, whereas more recently inundated sites were dominated by allochthonous carbon, with more than half the OM classified as recalcitrant. Disconnected systems therefore store more labile carbon, so that where human impacts modify conditions to favour mineralisation (such as through agricultural usage), the potential for floodplain carbon storage is limited (Figure 9.4). They argue therefore that full restoration of floodplains for carbon storage involves not only changes in connectivity, but reestablishment of more natural vegetation and land use on floodplains.

Hein et al. (2003), also working on tributaries of the Danube, demonstrated that these well-connected systems were dominated by allochthonous deposition at high flows, with autochthonous production contributing only 20% of total OM at mean

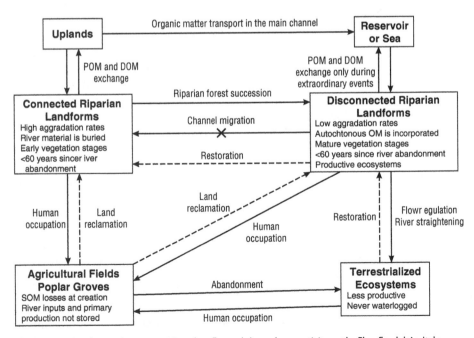

Figure 9.4   Landscape change, organic carbon flux and channel connectivity on the Ebro floodplain, Italy. Source: After Cabezas and Comín 2010. Reproduced with permission of Elsevier.

flows. During low flow periods, transformation of carbon dominated, with autochthonous planktonic sources of OM, becoming prevalent. The overall impact of restoration was shown to be an increase in total OM retention and transformation, and an increase of autochthonous POC export to the channel system. This pattern has also been identified in blackwater river systems in Florida. Here, restoration of the Kissimmee River channel through channel reconnection led to increased productivity to respiration ratios indicative of greater autochthonous production. Post-restoration, the reconnected floodplain was shown to be a key source of carbon to the channel (Colangelo 2007). DOC flux to rivers is rarely a river management criterion but Baldwin et al. (2016) and Stanley et al. (2012) have argued that the critical role of floodplains as carbon sources for river systems should explicitly acknowledge this role in the design of floodplain restoration schemes. The restoration of 'environmental flows' through enhanced connectivity would therefore be a primary target, with the aim of enhancing DOC flux to the river system to support fluvial food webs.

It is clear that across a range of river types and restoration approaches, the role of the floodplain in fluvial carbon cycling is central. Restoration of natural carbon dynamics that support the role of floodplains as a significant carbon store depend on two key elements. The first is restoration of natural flood inundation regimes, so that flows of sediment, water and nutrients between the channel and floodplain are effectively connected. Potentially of equal importance in the context of carbon sequestration is restoration of natural ecosystem functions in the riparian corridor. River restoration for carbon benefit cannot therefore be treated in isolation, but should be considered as part of a wider landscape restoration.

The contribution of geomorphological understanding to these approaches is important, firstly through the recognition that floodplain landforms are important carbon stores, and secondly through the accumulated geomorphological expertise in river restoration approaches, which can deliver geomorphological and hydrological reconnection of floodplains. However, this alone will not have significant carbon benefits in the face of intensive agricultural uses of floodplains. The caution from Dufour and Piégay (2009), that a return to pristine conditions is unachievable, is important in this context. For river restoration to deliver carbon benefits in intensively used landscapes, integration with wider conservation work or the development of lower intensity agricultural uses on floodplains may be required (Cabezas and Comín 2010). Carbon storage requires both the establishment of conditions for the growth and preservation of floodplain sediment stores, but also achieving biogeochemical conditions in the floodplain sediment store that are conducive to carbon preservation. This requires an interdisciplinary approach to river restoration and landscape management, with a focus on land use and ecosystem function, as well as physical connectivity.

Fryirs (2015) details ways in which geomorphological understanding has been incorporated into decisions on river rehabilitation, and suggests that restoration priorities are often driven by conservation principles, so that systems in good condition are prioritised because of the ecosystem services they provide. Sites in poor condition are low priority and intervention justified only when there are signs of natural regeneration of the system. Holl and Aide (2011) have similarly argued that natural regeneration is an undervalued resource in restoration work. However, in situations where geomorphological changes are leading to carbon loss from floodplain storage, either through erosion or through the impacts of reduced channel connectivity, appropriate intervention in reaches in poor condition may maximise ecosystem service benefits, since restoration entails an 'avoided loss' of carbon as well as the gains in carbon storage from intact systems.

Peatland Restoration, Geomorphology and Carbon Cycling

Peatlands are major carbon stores that globally sequester over 50% of soil carbon, an amount approximately equivalent to total atmospheric carbon storage (see Chapter 5). These substantial carbon stores have accumulated during the Holocene, but have come under very significant anthropogenic pressures over the past two centuries (Evans and Warburton 2007). Peatland degradation around the world has occurred due to overgrazing, wildfire and managed burning, pollution impacts, drainage, agricultural usage and resource extraction. Over the past two to three decades there has been an increasing recognition of the range of ecosystem services which peatlands provide, and so of the negative consequences of widespread peatland degradation.

Peatlands form at sites where the water table is consistently high, so that microbial decomposition of plant litter is inhibited due to low oxygen levels. High water tables are fundamental to healthy peatland ecosystems, and so the most widespread negative impact of human activity on peatlands is reductions in the water table with respect to the surface. Reestablishment of suitable water tables through blocking of drains, active water table management or management of peatland vegetation is a primary focus of peatland restoration (Bonn et al. 2016).

However, in cases of severe peatland degradation the initial focus of restoration efforts can be as much geomorphological as hydrological, since physical stability of the peatland surface is a prerequisite for restoration of active peat forming communities, which support peatland carbon sequestration. Examples of peatland erosion occur across the globe and include: severe erosion of blanket peatlands in the UK and Ireland (Evans and Warburton 2007; Tallis 1997); erosion of sloping percolation mires on the Ruoergai Plateau in China (Zhang et al. 2016); incision of valley mires in the Blue Mountains of Australia (Fryirs et al. 2016); and erosion of bare peat surfaces associated with peat mining which is common across northern hemisphere peatlands (e.g., Campbell et al. 2002; Kløve 1998).

Where physical erosion of peatlands has occurred, carbon dynamics are affected in three principle ways:

1) Through direct loss of particulate carbon (POC) from the peatland by erosion.
2) Through reduced net primary productivity (NPP) at erosional sites.
3) Through water table drawdown due to the drainage impact of gully erosion, which leads to higher rates of aerobic decomposition of near surface peats.

Analysis of the relative importance of these mechanisms in heavily gullied peatlands in the UK has shown that POC loss > reduced NPP > drainage impacts, with erosional POC losses the largest component (Table 9.1; Evans and Lindsay 2010b). In the most eroded systems, these carbon losses are sufficient to shift the peatland system from a net carbon sink to a carbon source (Evans et al. 2006).

A range of approaches to the restoration of eroding peatlands have been developed. Where bare peat is exposed by mining, typically on very low gradient sites, and where peatland erosion is predominantly wind driven, then management of water tables and re-vegetation either by moss transfer or natural re-vegetation may be sufficient to minimise surface erosion (e.g., Rochefort et al. 2003). However, where peat erosion has formed deep erosional gullies on higher gradient sites such as blanket peatlands, more active erosion control measures may be required to minimise particulate carbon losses from the peatlands.

Two approaches have been developed to stabilise degrading peatlands in the uplands of the UK. The first approach is re-vegetation of extensive areas of bare peat. The most severely eroded sites have large expanses of bare peat on interfluves as well as extensive gully erosion. These sites can produce very high levels of particulate carbon flux, up to circa 70 gC m^{-2} a^{-1} (Evans et al. 2014b; Pawson et al. 2012). Natural regeneration of these extensive bare peat surfaces is very poor because the frost disturbance and the acidity of the soil surface limit seedling establishment and survival (Anderson et al. 2009). The approach to re-vegetation that has been pioneered in the eroded peat landscapes of the south Pennines in the UK, involves aerial seeding with a nurse crop of utility grasses. Germination is supported by addition of lime (to raise the pH of the soil), fertiliser and the application of a mulch of cut heather (to promote seedling establishment) (Anderson et al. 2009). The approach leads to rapid establishment of a grass cover on the bare peat surfaces. The target restoration end-point for these systems includes re-establishment of native species (Figure 9.5).

The surface stability created by the nurse crop permits establishment of seedlings and these outcompete the nurse grass in the nutrient poor environment of the bog surface. The first native species to reestablish are typically heathers seeded from the cut mulch and significant cover of native species is achieved in 5–10 years (Alderson et al. 2019b). However, the benefits in terms of erosion control and POC flux are achieved much more rapidly. Vegetation cover of greater than 15–20% typically leads to significant reductions in surface erosion

Table 9.1   Relative contributions of three erosion related carbon loss mechanism to a 40 gC m^{-2} carbon loss estimated in gully eroded peatlands in northern England (Evans and Lindsay 2010b).

Mechanism	Proportion of carbon loss
POC loss	63–78%
Loss of NEE at eroded sites	17%
Reduced sequestration due to water table drawdown	4–5%

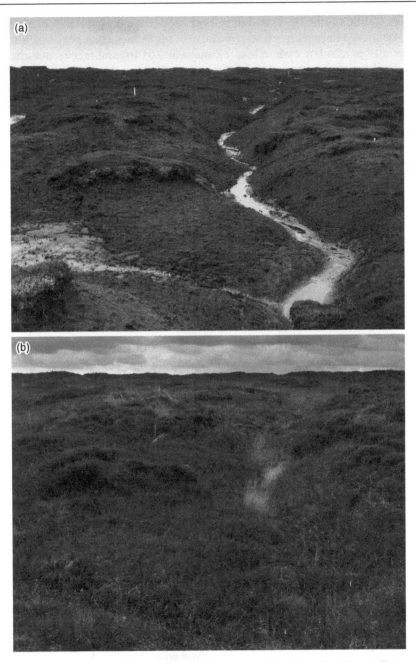

Figure 9.5   Before (a) and after (b) images of bare peat re-vegetated by aerial seeding. The period between the two images is circa 9 years. Source: Courtesy of Tim Allott.

(Rogers and Schumm 1991) and in re-vegetated peatlands, surface cover of circa 60% is achieved in two years and circa 80% in four years. Bare peat area is a good predictor of particulate carbon loss (Evans et al. 2014b) and recent measurement of particulate flux from eroding and restored sites has demonstrated that these rapid changes in surface cover produce more than an order of magnitude reductions in particulate carbon flux to levels comparable with intact peatland systems (Shuttleworth et al. 2015; Figure 9.6). The implication of these findings is that re-vegetation has the potential to restore carbon sequestration at degraded sites by dramatically reducing the largest carbon loss from the system.

A second approach to restoration in gullied systems has been gully blocking. Deep erosion gullies are a characteristic of the most eroded systems, and

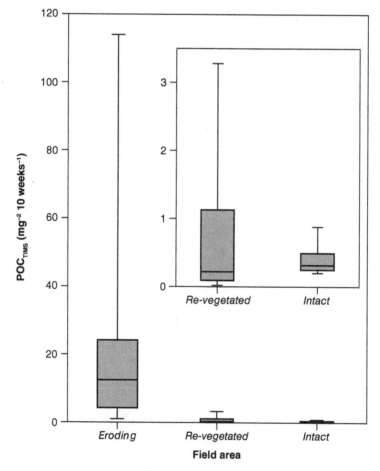

Figure 9.6   Impact of re-vegetation on particulate carbon flux from restored blanket peatlands. Source: After Shuttleworth et al. 2015. Reproduced with permission of Wiley.

approaches to blocking these gullies have developed from conservation practice (Brooks and Stoneman 1997, by analogy with drain blocking work (Ramchunder et al. 2009), and from work on blocking gullies in dryland and agricultural systems (Morgan 2009). From a geomorphological perspective, gully blocking aims to create zones of lower flow velocity promoting sediment accumulation and re-vegetation of the gully floor. This approach mimics natural re-vegetation of broad gully floors, where meandering flows create zones of lower flow velocity or where bank collapses create partial gully blockage (Crowe et al. 2008; Evans and Warburton 2007).

Gully blocks in narrow (1–2 m) width headwater gullies may be wood or plastic check dams spanning the whole channel (Figure 9.7a), which raise water tables close to the bog surface. These dams are analogous to drain blocking and hydrologically they have a positive impact on local water table. In broader gullies, stone or log dams create additional surface roughness in the channel and commonly fill with organic sediment creating low dams (Figure 9.7b). These blocks have demonstrated effective re-vegetation of gully systems (Figures 9.7b and 9.7c).

Gully blocking and re-vegetation restoration techniques are commonly used in combination. They represent the application of erosion control measures derived from applied geomorphological research, with the aim of enhancing the physical stability of the significant carbon store preserved in upland peatlands. Where particulate carbon losses dominate, the techniques have the potential to dramatically reduce carbon flux from peatland systems. However, it is important to note that they do not represent a complete restoration of these peatland systems. Where gully systems reach depths of 2–3 metres, gully blocks represent a stabilisation of the current morphology through reduction in erosion rates, but they are not a restoration of topography. The original pristine peatland is still significantly modified by dense drainage networks, which can substantially reduce water tables and enhance rates of peat decomposition.

Check dams or gully blocks have been used in other eroding peatlands globally. Zhang et al. (2016) report successful restoration of eroding Chinese peatlands using wooden and peat sack dams in combination with re-vegetation, leading to raised water tables. In common with the UK sites described above, the fast spreading wetland sedge *Eriophorum angustifolium* was an important species, which naturally recolonised areas rewetted by blocking.

The importance of geomorphological understanding in the management of upland peatlands has also been well studied in the context of upland valley bottom swamps in the Blue Mountains in Australia (Fryirs et al. 2016). In their natural condition, these sites are densely vegetated swamps characterised by high water tables and multiple discontinuous drainage lines. However, in a suburban environment many of these swamps have been subject to anthropogenic change, including clearance for grazing or development. Fryirs et al. assessed 458 sites and showed that 26% of sites were impacted, leading to channel incision into the valley fill. These sites have evolved towards a single thread channel form with

Figure 9.7   (a) Plastic piling gully blocks. Source: Martin Evans. (b) Stone dams at an eroded site before or shortly after gully blocking. (c) Eroded site 5 years later. (b) and (c) Source: Courtesy of Ed Lawrance.

knick points in the channel long profile. Fryirs et al. argue that increases in run-off flashiness due to catchment urbanisation, and localisation of flows through stormwater drains, are important in causing degradation of the swamp systems and changes in channel form. These geomorphological adjustments have also been shown to impact on the carbon storage potential of these systems. Cowley et al. (2016) have demonstrated that channel fills have significantly reduced organic matter content and C:N ratios below the level indicative of peat formation potential.

Understanding of the links between geomorphological form and carbon sequestration for these systems (Figure 9.8) is key to efforts to restore their carbon storage function. Freidman and Fryirs (2015) report on the efficacy of check dam structures in trapping sand and reducing gradients on channels draining incised peatlands, and argue that re-vegetation of these lower gradient surfaces is analagous to the early stages of swamp formation recorded in the alluvial stratigraphy. The rapid re-vegetation of check dam deposits is comparable to observations in gully-blocked systems in northern blanket bogs. These southern hemisphere valley peatlands (fens) are very different in detail to more widely studied ombrotrophic northern peatlands, but the fundamental approach to restoration of eroding systems, which is to reduce flow velocities and enhance sediment deposition to allow natural or artificial re-vegetation of the sites, is very similar.

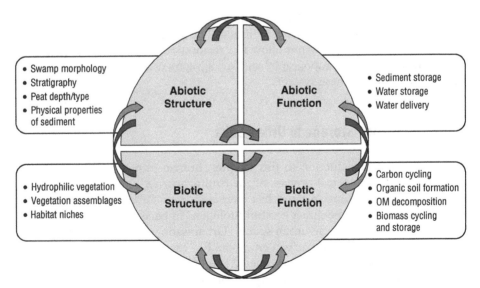

Figure 9.8   Links between Geomorphological and Biotic components of peatland ecosystems controlling carbon cycling in Australian upland swamp systems. Source: After Cowley et al. 2016. Reproduced with permission of Elsevier.

One of the consequences of erosion in valley bogs and in upland peatlands is the exposure of mineral sediment and an increase in mineral sediment flux through the system. This can produce a positive feedback, since the addition of mineral load to runoff waters has been demonstrated to accelerate rates of peat erosion by flowing water (Carling et al. 1997). Deposition of mineral material in the valley floors also produces significant zones of elevated hydraulic conductivity within the peat, as water drains preferentially through sand lenses within the stratigraphy (Fryirs 2015). These may lead to further drying of floodplain deposits and potentially to enhanced carbon losses. For both of these reasons, initiation of restoration works in eroding systems before significant exposure of mineral sediment to erosion is likely to be a preferred restoration strategy.

What is clear from the examples above, is that where peatland sites of carbon sequestration have become carbon sources through erosional processes, a geomorphologically informed approach to peatland restoration has the potential to deliver effective erosion control and to restore carbon sequestration functions to the peat system. Peatland restoration requires the establishment of the high water tables characteristic of intact systems and the establishment of suitable peat forming vegetation. Under these conditions, reduced decomposition and active accumulation of organic matter can restore the carbon storage potential, which is a key ecosystem service derived from peatland systems. Where there has been peatland erosion that modifies the natural topography of the peatland landform, geomorphological understanding becomes critical to restoration efforts. This includes both analysis of the drivers and mechanisms of erosion, so that suitable erosion control measures can be established. The scale of topographic modification of some peatlands may not allow full restoration, but an understanding of the impacts of modified topography and stratigraphy can guide optimal, achievable rehabilitation of these systems.

Managing Carbon Storage in Urban Soils

As noted in the introduction to this chapter, human impact is potentially the defining geomorphological process of the Anthropocene. The proportion of the world's population living in cities has increased from 30% to 54% since 1950 (UN 2014), and thus the direct geomorphological impacts of human activity are particularly intense in these urban spaces. Urbanisation has a significant impact on carbon cycling within the urban envelope. For example, rapid urbanisation of Shenzhen in China has led to an estimated loss of NPP of 1.7 gC m^{-2} a^{-1} (Deyong et al. 2009; Rawlins et al. 2015). However, cities are inherently managed environments and so there is the potential to modify human actions so as to minimise carbon losses to the atmosphere. The potential to manage urban systems for environmental gain is central to ideas of sustainable urbanisation (UN 2014), and one potential area for intervention is consideration of carbon storage in urban sediments.

'Made soils' in urban environments have the potential to store as much carbon as natural systems, and in drier parts of the world considerably more (Pouyat et al. 2002, 2006). In particular, the addition of organic wastes as soil improvers has the potential to produce soil carbon storage in urban soils of 22 gC m^{-2} a^{-1} (an example from a study of soils in Tacoma, Washington; Brown et al. 2012). Pouyat et al. (2002) identify soil stability as a key control on the carbon content of urban soils. Highly disturbed soils have much lower carbon contents than natural systems, whereas the soils beneath suburban lawns are comparable to grassland soils more widely. Similarly, Trammell and Carreiro (2012) have modelled the impact of soil disturbance associated with highway construction in Kentucky and showed that the disturbed systems had 3% less soil carbon, but 20% less tree carbon. However, the extent of these verge environments may mean that, suitably managed, they are potential carbon stores. Mestdagh et al. (2005) have estimated that the soil carbon stock associated with grassy verges in Flanders is equivalent to 10% of the total regional grassland stock.

Because stability appears to be a key control, preservation of established green space is required to maximise carbon storage in urban soils. One unlikely but potentially important source of stability in these systems is the addition of impervious surfaces. Edmondson et al. (2012) studied a complete organic carbon stock across the city of Leicester in the UK, and report that organic carbon storage in 'capped' soils below impervious surfaces was equivalent to that observed in urban greenspaces and that the stocks exceeded local agricultural land (Figure 9.9). It seems likely therefore that both made ground and intact soils sealed beneath urban infrastructure are potentially important soil carbon stores in the urban context.

Carbon sequestration by sediments in urban systems can occur in novel ways. For example, urban soils have been shown to be significant stores of black carbon produced by combustion of fossil fuels. Black carbon is highly recalcitrant and so black carbon-rich sediments are potentially a stable long-term carbon store. Edmondson et al. (2015) showed high rates of carbon storage in urban soils in northeast England. Here, soil carbon stocks of 31–65 kgC m^{-2} exceeded regional forest soils, but up to 39% of that stock was black carbon. Black carbon was enriched under trees, consistent with the importance of vegetation-filtering effects in capturing atmospheric black carbon to soil storage (Brantley et al. 2014).

Urban soils may also capture carbon directly from the atmosphere. Significant amounts of made ground in urban areas is produced by recycling cementitious demolition waste. Washbourne et al. (2012) have demonstrated that weathering of these soils can lead to secondary carbonate deposition and they can store significant amounts of carbon drawn down from the atmosphere. This is analogous to the formation of pedogenic carbonate in areas of carbonate rock weathering, but is driven here by the addition of high calcium silicates from building waste to soils. Washbourne et al. estimate that at their site up to 3.4 kgC yr^{-1} can be sequestered per tonne of soil.

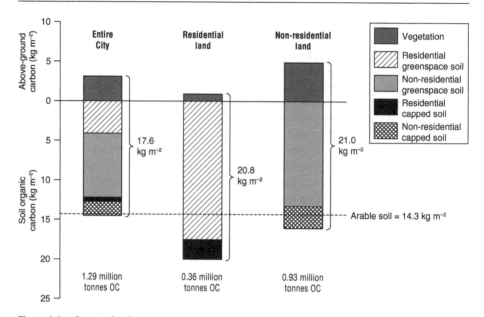

Figure 9.9 Estimated carbon storage in Leicester, UK. Source: After Edmondson et al. 2012. Reproduced with permission of Springer Nature.

It is clear through the examples above that there is potential for active management of urban soils and sediments to promote carbon sequestration. Urban environments are Anthropocene landscapes where ecosystem services are controlled by human activity, and so provide opportunities for the deliberate manipulation of urban carbon sequestration in these contexts.

Urban geomorphology is a rapidly developing interdisciplinary field (Ashmore and Dodson 2017; Thornbush 2015), which draws on work on urban rivers (Gurnell et al. 2007), weathering of building materials (Pope et al. 2002), constructed landforms (Csima 2010, Goudie and Viles 2016) and the development of unique urban landforms (Dixon et al. 2018), as well as understanding the geomorphological context for development work (Cooke 1976).

To date, explicit applications of geomorphological thought to questions around urban carbon cycling and storage have been limited. As discussed in the previous section, the application of geomorphological expertise in river restoration in the urban context can make a major contribution here. However, the observation of the importance of urban soils and sediments as potentially important sites of carbon sequestration also focusses attention on geomorphological understanding of sediment flux and sediment storage in urban contexts. Critical in assessing the role of urban systems in the terrestrial carbon cycle is quantification of the magnitude of made ground in the urban system and assessment of the stability of these stores. The work of Douglas and Lawson (2000), in estimating human impacts on geomorphic work at landscape scales, illustrates a potential way forward through the application of sediment budget methodologies (Dietrich et al. 1982; Walling and Collins 2008) to anthropogenic modification of urban surfaces. However, sediment budgets alone will not deliver a full understanding of terrestrial carbon cycling.

The importance of sediment store stability identified in the studies above requires consideration of a further element of urban geomorphology, namely the rethinking of disturbance regimes in the urban context. Geomorphologists have a good understanding of return periods of natural drivers of disturbance of sediment storage, including storm-induced fluvial erosion and mass movements. In some urban contexts these can be relevant, but understanding disturbance of urban carbon stores will also require acknowledgement of human drivers in urban geomorphology. Relevant timescales for the disturbance of urban sediments may be driven by economic cycles, and the lifespan of urban fabric, which drive rates of construction and redevelopment. Understanding the drivers of urban geomorphology may require looking beyond the natural sciences to interdisciplinary collaboration with the social sciences.

The final element of the urban carbon storage system is understanding of the processes of carbon cycling in the urban context. This requires further work on the organic matter quality of urban sediments and soils, to understand sources of organic carbon (e.g., black carbon vs. autochthonous OM production) and requires opening up the black box of bulk sediment carbon determinations to gather insight into the physical and microbial processes which control carbon turnover in these systems. This is another area where more research is required. Urban soils have been shown to support higher microbial biomass than rural systems in Colorado, leading to higher measured rates of carbon turnover (Kaye et al. 2005) and the importance of disturbance to microbial activity has been demonstrated for urban soils in Beijing (Wang et al. 2011). In a study of construction work in Virginia, USA, construction disturbance has been shown to significantly reduce soil carbon content, although urban soil rehabilitation through organic matter addition appears to have enhanced carbon storage below the top layers of the soil (Chen et al. 2013a). This study attributed this storage to enhanced aggregate formation associated with elevated microbial biomass in the restored soils, a pattern which has also been reported for urban soils in Germany (Beyer et al. 1995). Progress in understanding carbon sequestration and loss in urban sediments will require synthesis of this type of process understanding, with analysis of sediment residence times controlled by natural and anthropogenic erosion and deposition regimes.

Conclusions

Recognition that human impacts are a dominant driver of ecosystem change is a threat, but also an opportunity. Understanding the role of human activity in changing ecosystems and the ability of human action to effect significant changes implies that there are options to mitigate the negative impacts of anthropogenic change. This chapter argues that understanding the role of geomorphology in the terrestrial carbon cycle is central to designing and implementing effective carbon management strategies in a variety of landscape systems.

Total global soil organic carbon storage is estimated at 1460 PgC (range 500–3000) (Scharlemann et al 2014). Floodplain carbon storage is estimated at 0.5–8% of this value (8–13 PgC, Sutfin et al. 2016) and peatland storage as 31% of the total (450 PgC, Gorham 1991). These are figures that compare to the atmospheric carbon store, which is equivalent to around 55% of the soil carbon storage (829 PgC, Cubasch et al. 2013). Therefore, whilst the carbon impact of local peatland or river restoration may be relatively small, the landscape scale effect of the 'avoided carbon loss' derived from restoring these widely degraded systems to good ecological condition is potentially large. Carbon sequestration at timescales appropriate for long-term mitigation of climate change requires long-term storage of the carbon captured from the atmosphere. In the terrestrial context, this means storage of carbon in soils and sediments. Three elements of geomorphological understanding are relevant:

1) *Identification and quantification of carbon storing landforms* including depositional (e.g., floodplains) and constructional (e.g., manmade levees) landforms. Relevant approaches include sedimentology, geomorphological mapping and digital terrain analysis (Chartin et al. 2011).

2) *Analysis of carbon store stability.* Rates of carbon storage and remobilisation in depositional landforms can be understood through geomorphological approaches such as sediment budgeting and interpretation of Quaternary stratigraphy. Applied geomorphological understanding of erosion control techniques provides the opportunity to enhance the physical stability of key carbon stores. Concepts such as sediment connectivity are valuable in understanding carbon fluxes between storage zones and in developing interventions in the particulate carbon cycle.

3) *Understanding of sediment sourcing and sediment character* through approaches such as sediment fingerprinting are an important input into analysis of the character and consequently the stability of organic matter in carbon storage zones.

Taken together, these characterisations of the physical form, history and stability of depositional landforms where carbon is stored, describe the environmental context and physical stability of carbon storage. A complete understanding of carbon dynamics requires an interdisciplinary approach, which also characterises the microbial and chemical processes that underpin carbon transformation and release, and the ecological processes that control autochthonous carbon production. Understanding of the interactions between physical setting and microbial carbon turnover is relatively limited, and this is an important research frontier in restoration science. Most restoration at present establishes physical stability, and may target other physico-chemical restoration targets such as water table or water quality, but proper linkage of landform and carbon turnover processes is a significant research challenge. More progress has been made in understanding the reciprocal relationship between biota and landform (Crowe et al. 2008), but again the linkage to carbon cycling is underexplored. Progress in these key areas will be central to transforming anthropogenic impacts on the terrestrial carbon cycle, from atmospheric carbon production, to mitigation of the impacts of rising atmospheric carbon concentrations.

Chapter Ten
Towards a Geomorphologically Informed Model of Terrestrial Carbon Cycling

Introduction

The carbon landsystem models developed in Chapters 5 to 8 describe the physical context of terrestrial organic carbon dynamics across a range of landscape types through the sediment cascade. It is clear from the review of empirical understanding of the terrestrial carbon system that a major challenge in understanding these systems is the integration of biological and sedimentary dynamics (Figure 10.1), operating at diverse time and space scales.

The challenge of upscaling process understanding is a familiar task for geomorphologists, who have long sought to understand large-scale landform and landscape change, through the means of small-scale measurement, and the explanation of processes and rates of erosion and deposition. The preceding chapters have highlighted ways in which geomorphological understanding of sedimentary dynamics can contribute to a more complete analysis of the terrestrial carbon system. This chapter will aim to build on this understanding of the carbon–sediment cascade and elucidate geomorphological approaches to understanding landscape sediment flux (as considered in Chapter 4), to develop an approach to the analysis of terrestrial carbon cycling, which can accommodate the range of relevant scales and processes involved.

Geomorphology and the Carbon Cycle, First Edition. Martin Evans.
© 2022 Royal Geographical Society (with the Institute of British Geographers). Published 2022 by John Wiley & Sons Ltd.

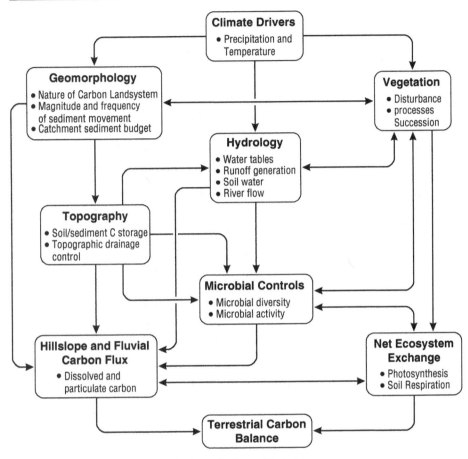

Figure 10.1   Controls on the terrestrial carbon cycle. Source: Martin Evans.

Controls on Carbon Sequestration in the Terrestrial System

The net carbon flux at a point in the landscape can be defined as Equation 10.1:

Equation 10.1: $Net_C = (NPP - R) + S - E + DOC_{in} + DOC_{out} + DIC_{in} - DIC_w$

where Net_C is net carbon flux at a point in the landscape, NPP is net primary productivity, R is gaseous carbon losses related to decomposition of organic matter (equivalent to heterotrophic soil respiration), S is carbon added to the system by sedimentation, E is carbon removed from the system by erosion, DOC_{in} and DIC_{in} are additions of dissolved organic and inorganic carbon respectively from upstream/upslope, and DIC_w is inorganic carbon lost from the system through weathering.

The first term of this equation defines the balance between production and decomposition of organic matter, which is controlled at an ecosystem level by vegetation type (species and growth form) and light conditions (which drive photosynthetic fixation of carbon) and by microbial decomposition of plant litter and soil organic matter. The S and E terms represent geomorphological processes, driving sediment flux through the system. The DOC and DIC terms relate to inputs and outputs of dissolved carbon (organic and inorganic respectively). DOC and DIC inputs are related to the flux of dissolved carbon from upstream or uphill. DOC outputs are a function of inputs, DOC mineralisation on-site and DOC production on-site, driven by rates of OM decomposition. DIC outputs are a function of on-site weathering and OM mineralisation.

A key generalisation that can be made from the preceding review of carbon cycling through the sediment cascade, is that biological and sedimentary elements of the terrestrial carbon cycle may be operating at divergent timescales, where timescales in the order of centuries to millennia may be required to achieve equilibrium sediment flux under geomorphological control (Church and Slaymaker 1989; Hoffmann 2015). This timescale is distinct from the decadal to century timescales required to achieve equilibrium in vegetation systems.

Long-term (years to millennia) carbon sequestration at a point in space is a function of primary productivity (vegetation controlled), decomposition rates (microbially controlled) and geomorphological controls on the persistence of sites of terrestrial carbon storage (landforms). Figure 10.2, based on the discussion in Chapters 5–8, is a generalisation of the sediment cascade identifying major carbon stores and fluxes of carbon through the system. Fluxes are classified according to which of these three drivers is most important as a control on flux. Although characteristic timescales of change may vary between these three main drivers of carbon sequestration and turnover, there are also important feedbacks between them.

Microbial communities respond in the short term (days to weeks) to changes in moisture and temperature, but are also affected by changes in vegetation type, changes in litter quality, the rate of photosynthesis and nutrient allocation within plants (which modify production of root exudates and litter quantity) and the impact of plants on soil moisture regime (Bardgett et al. 2007; Gutknecht et al. 2012; Sayer et al. 2017). All of these factors are important vegetation-driven feedbacks to microbial growth and function. Bardgett et al. (2007) suggest that whilst the assimilation of carbon is well understood, soil carbon losses through respiration are more poorly described because of the complexity of below-ground ecosystems. Sayer et al. 2017 argue that the interdependence of plant and microbial ecosystem components is central to ecosystem function, and is a key control on plant succession, and so on ecosystem recovery from disturbance. Therefore, long-term change in vegetation communities will drive changes in delivery of carbon substrate for decomposition by microbial systems. Similarly, the delivery of nutrients to plants through microbial decomposition is also a significant control

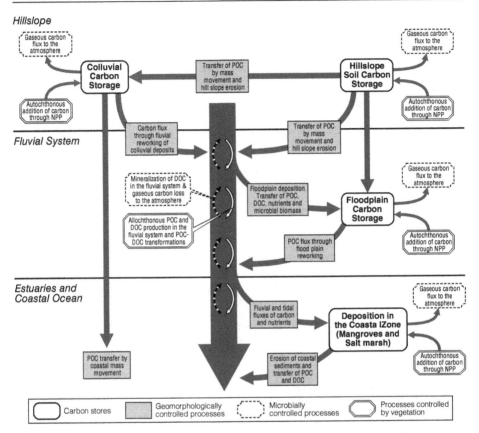

Figure 10.2   Key stores and fluxes of carbon in the sediment cascade. Dominant controls classified as Vegetation, Microbial or Geomorphological in the legend. Source: Martin Evans.

on vegetation change (e.g., as exemplified by succession on glacial forefields discussed in Chapter 5).

In addition to the interlinkages between vegetation and microbial communities, geomorphological change also plays a major role in terrestrial carbon cycling, as has been demonstrated in preceding chapters, both through the movement of particulate carbon around the landscape by processes of erosion and deposition, but also through the modification of the geomorphology and sedimentology of sites of carbon sequestration. These changes impact on water tables, microbial niches, litter quality and can drive successional change in both microbial and plant communities.

Figure 10.3 describes the key drivers of terrestrial carbon cycling from Figure 10.2 across timescales from days to millennia. As noted above, the characteristic timescales of change vary between the three key drivers of the terrestrial car-

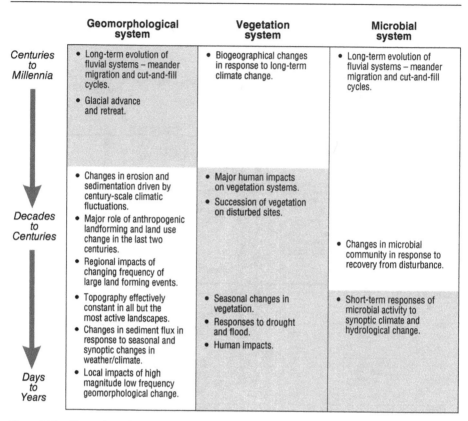

	Geomorphological system	Vegetation system	Microbial system
Centuries to Millennia	• Long-term evolution of fluvial systems – meander migration and cut-and-fill cycles. • Glacial advance and retreat.	• Biogeographical changes in response to long-term climate change.	• Long-term evolution of fluvial systems – meander migration and cut-and-fill cycles.
Decades to Centuries	• Changes in erosion and sedimentation driven by century-scale climatic fluctuations. • Major role of anthropogenic landforming and land use change in the last two centuries. • Regional impacts of changing frequency of large land forming events.	• Major human impacts on vegetation systems. • Succession of vegetation on disturbed sites.	• Changes in microbial community in response to recovery from disturbance.
Days to Years	• Topography effectively constant in all but the most active landscapes. • Changes in sediment flux in response to seasonal and synoptic changes in weather/climate. • Local impacts of high magnitude low frequency geomorphological change.	• Seasonal changes in vegetation. • Responses to drought and flood. • Human impacts.	• Short-term responses of microbial activity to synoptic climate and hydrological change.

Figure 10.3   Timescales and drivers of change in geomorphological and biological systems. Source: Martin Evans.

bon system. One of the seminal contributions of twentieth-century geomorphology was a paper by Schumm and Lichty (1965), which considered 'Time Space and Causality in Geomorphology'. This paper examined the ways in which landforms develop in response to geomorphological processes operating at diverse timescales, from the minutes to hours of storm events, through to geological timescales controlling tectonic processes. The paper hypothesises that at short timescales, geomorphic systems can be regarded as in steady state (dynamic equilibrium) with many controls on the system essentially static boundary conditions (independent variables) at these timescales. At longer timescales a wider range of dynamic dependent variables control landform evolution. This conceptualisation can usefully be applied to divergent timescales controlling carbon cycling within the sediment cascade.

At timescales of days, the primary control on carbon sequestration is the balance of primary productivity and microbial decomposition of organic matter, which at this timescale is likely driven by short-term variations in photosynthesis linked to light levels, and variation in microbial activity in response to changes in temperature, moisture and nutrient regimes. Also relevant at these timescales are synoptically-driven fluxes of organic sediment. Therefore, at timescales of days, the macro-scale geomorphology of the landform and the vegetation cover are independent variables in the terrestrial carbon cycle, with synoptic changes in hydrometeorology and microbial responses representing the key dependent variables.

At timescales of years to decades, changes in vegetation cover have the potential to significantly alter primary productivity at a site, and so to shift the productivity–decomposition balance. At these timescales, vegetation changes are commonly a response to disturbance that could be either natural or anthropogenic. Therefore, biological processes of succession and the associated changes in vegetation and microbial communities are key controls on carbon cycling, and become dependent variables at these timescales.

At timescales of centuries to millennia, rates of carbon storage in carbon sequestering landforms potentially becomes a less significant control on long-term carbon balances than the persistence of the landforms in the landscape. Large-scale geomorphological processes, such as meander migration or glacial advance and retreat, become dependent variables, and may drive large-scale lateral transfers of carbon within the sediment cascade. Centuries or millennia of carbon storage can be rapidly removed by erosion so that it is erosion rates, and consequently, residence times of carbon in carbon storing landforms which become the dominant control on the local carbon cycle. By comparison to biologically focussed work on ecosystem carbon cycling, analysis of such systems requires the integration of the longer timescales associated with landscape change and a recognition that organic carbon fluxes through the terrestrial system are non-stationary (Kuhn et al. 2009).

In essence, the argument above is that over the shortest timescales (up to a decade) variability in the biological system exceeds that in the geomorphological system. At longer timescales, the magnitude of changes in the terrestrial carbon balance at a point in space (which can be triggered by geomorphological change) exceed the variability of the biological system. Arguably, it is in the intermediate timescales of decades to centuries where understanding the interaction of biological and geomorphological change is most urgent, not least because these are timescales of human experience, and the timescales over which the earth is predicted to experience significant climate change.

Changes in boundary conditions such as progressive climate change produce non-equilibrium conditions. Implicit in the discussion above is that biological systems are likely to achieve equilibrium conditions post-disturbance more rapidly than the geomorphological system. In non-equilibrium conditions, the relative magnitude of net ecosystem exchange and erosional losses of carbon are significantly disrupted. In systems where long-term erosion of carbon stores is occurring, then carbon sequestering landforms, which may have accumulated carbon over millennia, have the potential to become significant sources of carbon to the atmosphere. An example of this process is the erosion of upland peatlands. Intact upland peatlands sequester on the order of 50 gC m^{-2} a^{-1}, whereas rates of particulate carbon loss from eroding systems can be hundreds of gC m^{-2} a^{-1} (Evans and Warburton 2007). The crossing of a geomorphological threshold and initiation of erosion (in the case of upland peatlands often triggered by human impacts on the vegetation system) creates disequilibrium conditions, in which geomorphological controls on carbon cycling dominate the carbon budget of the sediment store. A similar impact of disequilibrium conditions generated by erosion of a significant carbon store is occurring in coastal systems, where loss of mangrove and salt marsh through erosion related to sea level rise exceeds rates of *in-situ* carbon sequestration.

Connectivity and Transit Times Through the Sediment Cascade

Hoffmann (2015) has demonstrated that highly connected sediment cascades have faster sediment transit times, and thus reduced residence time of sedimentary material within depositional landforms. Very short residence times (hours to months) are likely to limit carbon turnover within the sediment system, because the time available is less than that required for biological decomposition of particulate organic matter. For example, in very short, steep river systems (e.g Waipoa River New Zealand; Leithold et al. 2016), high connectivity and short transit times lead to the export of large amounts of particulate carbon to the oceans. It is likely that moderate connectivity drives more rapid carbon turnover than very low connectivity. The critical importance of microbial niches in driving decomposition (Schmidt et al. 2011) means that disturbance and oxygenation of the sediment in a more dynamic system (rapid turnover) will drive more immediate carbon mineralisation. In contrast, in poorly connected systems in conditions where decomposition is supressed, and where long residence times are characteristic, more carbon will be stored (Figure 10.4).

A consequence of the hypothesis outlined in Figure 10.4 is that the impact of erosion on the carbon cycle depends on where the erosion is located; for example, hillslope erosion on poorly connected hillslopes will drive carbon storage in footslope deposits, whereas actively eroding fluvial systems that are reworking floodplain and fan deposits are likely to promote disturbance and carbon loss from both stores and within the fluvial system itself. The mathematical framework for analysis of the sediment cascade proposed by Hoffmann (2015) has significant potential value in assessing the interactions of sedimentary and biological processes within the sediment cascade. Knowledge of characteristic organic matter residence times within depositional landforms defines an appropriate time base over which to extrapolate gaseous carbon losses, in order to produce meaningful carbon budgets.

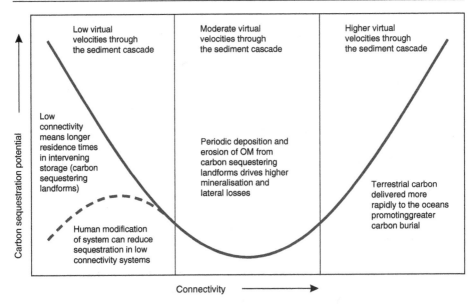

Figure 10.4   Hypothetical relationship between sediment cascade connectivity and landscape carbon sequestration potential. Note the main curve refers to unmodified systems. As discussed in Chapter 7, in relation to floodplain modification of unconnected systems, this can reduce carbon sequestration potential. Source: Martin Evans.

A limitation of much of the work considering sedimentary controls on terrestrial carbon cycling is a focus simply on the residual of the carbon budget (carbon sequestration). Carbon storage in depositional landforms is the residual of carbon deposited at the site, carbon produced by *in-situ* NPP and gaseous and dissolved carbon losses resulting from microbial decomposition. Gaseous loss can be estimated if the other parameters are known, but as has been discussed in previous chapters, the challenge is ensuring that estimates of carbon input (based on erosion rates) and carbon storage are temporally representative. The alternative approach to assessing gaseous losses is direct measurement of fluxes at depositional sites (e.g., Van Hemelryck et al. 2011). This allows assessment of gaseous losses directly, but the short-term measurements have to be extrapolated to understand longer-term carbon budgets, and short-term measurements may not be representative of longer-term change, particularly where disturbance drives successional change in vegetation.

The evidence that organic matter decomposition in depositional sites may continue over timescales of centuries to millennia (Wang et al. 2010) means that estimation of a representative gaseous carbon flux comparable to observed sediment storage requires evidence of the time period of deposition at the site. Defining reservoir storage terms (Hoffman 2015) and hence characteristic sediment transit times through the sediment cascade, offers the potential for direct linkage of biological measurements of carbon flux at storage sites, to understanding of the sediment dynamics of those sites at appropriate timescales. A characteristic of the sediment cascade is the nested nature of controls on sediment flux at increasing spatial scales (Slaymaker et al. 2009). A standard assumption of geomorphological analysis of landscapes is that time and space scales are linked, so that to explain landforms and sediment fluxes in larger basins also requires consideration of longer timescales.

Figure 10.5 is based on an illustration of this concept presented by Slaymaker et al. (2009), modified to consider the major controls on carbon sequestration across a range of spatial and temporal scales within the sediment cascade. The virtual velocity of sediment transfer through the cascade is illustrated as a vector through this time–space plot. In Figure 10.5, this is a linear vector, which assumes that virtual velocities are constant with scale. However, increases in accommodation space and reductions in gradient through the cascade may lead to variable virtual velocities. Therefore, since landform stability and carbon storage times are key controls on carbon flux through the sediment cascade, understanding the role of geomorphological processes in catchment carbon fluxes requires analysis of landform stability at a range of scales.

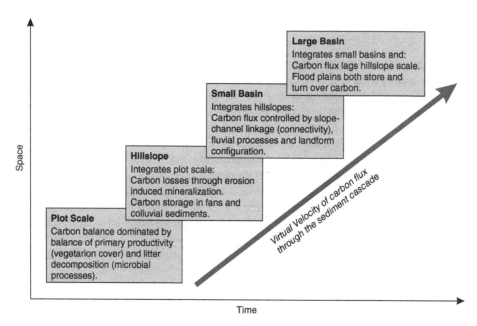

Figure 10.5 Controls on carbon flux through the sediment cascade at increasing time and space scales. Source: Martin Evans.

A Geomorphological Approach to Terrestrial Carbon Cycling

A strongly empirical focus on processes and landforms has meant that the development of conceptual models as a basis for further empirical work, which is common in the ecological literature, is not a standard approach in geomorphological research. The relative lack of overarching conceptual frameworks means that generalisation from the large quantity of first-class empirical research that underpins understanding of local landscape systems is more challenging. As has been argued by Slaymaker and Spencer (1998), a contribution to understanding of biogeochemical cycling should be an important focus for geomorphological research, which in combination with standard geomorphological approaches has the potential to serve as an integrating conceptual framework for diverse fields of study. In the context of increasing concern about global environmental change, this focus is even more important.

Geomorphology as a discipline has the aim of understanding the ways in which erosion and deposition combine to produce landforms, and furthermore the ways in which landforms combine to define landscapes. Inherent in this definition is the challenge of explaining landscape scale phenomena through the upscaling of process understanding at the scale of the landform or below. Consequently, as has been discussed above, the one area where geomorphology *can* claim to have developed useful theoretical knowledge is in approaches to integrating understanding of small-scale processes into analysis at landscape scales. Issues of scaling are central to the challenge of integrating an understanding of lateral carbon fluxes into models of the global terrestrial carbon cycle, and this section explores ways in which geomorphological ideas and approaches can be integrated into a conceptual analysis of the system.

The starting point for this analysis, from the discussion above, is the notion of landscape stability. Depositional landforms are sites of significant carbon sequestration, which can lock up terrestrial carbon for periods of centuries to millennia. Application of chronostratigraphic techniques to understand the typical age of depositional features means that we can estimate average rates of carbon sequestration, and also defines the maximum timescales for carbon turnover. However, the eventual carbon storage at a depositional site, derived from calculation of sediment volumes and organic content, and from dating of sediment sequences, is only a partial description of their role in the carbon cycle. Depositional landforms are not simply sites of passive carbon accumulation, but they are dynamic components of the terrestrial carbon cycle. Current stocks of carbon are a product of both allochthonous and autochthonous carbon sources to the landform and of the rates of carbon mineralisation within the sediment system. Carbon is also lost from depositional sites by erosion and disturbance regimes have significant influence on NPP and vegetation patterns, and eventually on soil development.

This book has explored the ways in which geomorphological understanding can contribute to a fuller understanding of the processes controlling fluxes of terrestrial carbon. In the following sections, four key aspects of terrestrial carbon storage in soil and sediment systems are discussed, and the basis for a geomorphological approach to their analysis is outlined.

The Relation of Erosional Disturbance and On-Site Primary Productivity

As outlined in Chapter 6, on hillslope sites where erosion is occurring, a key factor in controlling the ecosystem carbon balance is the magnitude and return period of the erosional disturbance. In natural systems, the recovery of NPP post-disturbance is generally rapid. Weng et al. (2012) modelled the response of ecosystem carbon storage to disturbance, demonstrating that recovery from intermittent disturbance is rapid and leads to minimal reduction in overall carbon storage. This is consistent with the observation that rapid re-vegetation of landslide scars in steeplands can lead to a net increase in carbon stocks (Page et al. 2004). In contrast, in agricultural systems where the disturbance regime is an annual cropping cycle, maintenance of NPP is dependent on enhancement through artificial fertilisers. Weng et al. (2012) emphasise the importance of considering carbon balance at appropriate spatial scales, since measurement of significant local impacts in a spatially diffuse disturbance regime may not imply major changes in the overall carbon budget at the landscape scale (e.g., Kashian et al. 2006). Running (2008) has argued that the next generation of global climate models should include the impacts of disturbance, including erosion linked to agricultural modification of ecosystems, wildfire and disease. The evidence that geomorphological change, such as disturbance due to typhoon driven landsliding (Lin et al. 2008), impacts significantly on local carbon cycling, means that this argument should also be extended to consider ecosystem disturbance linked to hillslope instability.

The need for geomorphological understanding: Controls on landslide occurrence and return period are central to geomorphological research on hillslope sediment transfer. Recent work has emphasised both the role of landsliding in modifying hillslope net ecosystem exchange and in export of POC to the fluvial system (Hilton et al. 2008, 2011b). This work considers both the role of geomorphological processes in carbon cycling, but also the control of geomorphological factors such as relief on carbon fluxes generated from steepland systems. Fuller integration of these approaches with biogeomorphological research on links between disturbance regimes and variation of NPP in space (e.g., Schimel et al. 1997) has the potential to develop relations between landscape and landform and an understanding of local carbon balance, which includes both successional and geomorphological forcing. Quantifying the topographic drivers of these processes offers the potential to model them at the landscape scale.

Carbon Turnover Potential in Depositional Landforms

Traditional approaches to investigate the preservation of organic carbon in soils has distinguished distinct carbon pools which turn over at different timescales. This conceptualisation underpins most models of soil carbon cycling (McGill 1996). However, as discussed in Chapter 6, it has more recently been argued that all soil organic matter is subject to rapid decomposition under suitable conditions and over timescales on the order of 50 years (Schmidt et al. 2011). This paradigm shift has important consequences for understanding of carbon cycling through the sediment system. Whereas approaches such as the river continuum concept (Vannote et al. 1980) assume an increasing age and recalcitrance of OM as it progresses through the sediment cascade, a focus on the conditions for decomposition means that the nature and stability of depositional sites becomes the key control on where carbon is effectively sequestered. In the soil system, the focus shifts from the presence or absence of suitable microbial decomposers for particular carbon pools to suitable conditions for the activity of relatively ubiquitous decomposer species. Two major research requirements emerge from this analysis. The first is the need to establish the relative importance of presence vs. activity in microbial decomposer communities. Key to understanding processes of decomposition at a particular site is not simply which decomposers are present (determined through DNA analysis), but which are active (through analysis of RNA). This will develop understanding of links between sediment conditions and microbial activity. The second research need is to understand the relations between landform, topographic position, and the soil/sediment environment, which combine to support decomposer activity. Modelling of soil saturation and organic matter content using topographic wetness indices (e.g., Pei et al. 2010) has the potential to define patterns of soil condition conducive to mineralisation/sequestration dominance. Therefore, geomorphological mapping of carbon sequestering landforms, along with an understanding of where in the landscape these occur, can define landscape scale carbon storage potential. By understanding the magnitude, location and character of carbon sequestering landforms, in addition to the potential for *in-situ* decomposition of stored carbon, it should be possible to establish an ecosystem scale propensity to sequester carbon.

The need for geomorphological understanding: The geomorphological contribution to this work includes both the geomorphological expertise required to identify and quantify carbon sequestering landforms, and research which seeks to understand topographic drivers of ecosystem processes. Empirical work is required to establish patterns of carbon storage for carbon landsystems across the sediment cascade, under varying conditions of climate and land-use. Critically, this understanding extends beyond the identification of zones of carbon storage, to gathering insight into the stability of stored carbon in a variety of landscape contexts. This requires collaboration between microbiologists developing understanding of rates of decomposer activity, geomorphologists characterising carbon stores, and terrain analysts exploring where in the sediment cascade stable carbon storage can occur.

Characteristic Timescales for Physical Stability of Depositional Landforms

Landscape scale connectivity of the sediment system occurs at timescales from decades to millennia. The intermittent linkage, particularly between hillslopes and channels and of channels and floodplains, is a critical determinant of rates of POC flux through landscape systems (Galy et al. 2015). The long-term equilibrium carbon storage potential of a landform is a function of the rate of carbon sequestration at the surface, and the balance of erosion and deposition at the site. The average persistence of a landform in the landscape is therefore also a fundamental control on net rates of carbon sequestration. Geomorphological modelling approaches validated through empirical understanding of the chronology of erosional and depositional phases are required to define characteristic timescales over *which* rates of particulate carbon flux through the sediment cascade (and net carbon storage) can be regarded as in equilibrium.

The need for geomorphological understanding: Estimates of characteristic timescales for the remobilisation of stored carbon may be derived from modelling approaches. For example, models of meander migration (e.g., Coulthard and van de Wiel 2006) offer the potential to estimate timescales for complete reworking of sediment storage at a river cross-section. However, such estimates require validation through geomorphological and geochronological approaches to dating depositional landforms. Extensive data already exists within the geomorphological literature to explore the persistence of carbon stores in the landscape. The extensive literature on the Quaternary history of fluvial systems has generated large datasets of radiocarbon dated fluvial deposits (e.g., Lewin et al. 2005) and similar data exist in relation to the age of colluvial fan deposits (e.g., Chiverrell and Jakob 2013) and a wide range of other sedimentological contexts. The work of Hoffmann (2015) discussed above proposes one approach to integrating these data into a model of sediment and carbon fluxes through the sediment cascade.

Variability of Sediment Cascade Connectivity in Space and Time

Understanding changes in connectivity within the sediment system driven by intrinsic change may define the timescales at which reasonable long-term average carbon flux can be defined. For example, rates of carbon storage in floodplains which grow by vertical accretion may decline as increases in floodplain height reduce inundation frequency, and lateral reworking of floodplain deposits becomes a relatively more significant process. Average floodplain storage of carbon should therefore be calculated at timescales in excess of the time period of this intrinsic change. However, long-term variability in sediment and carbon flux through the sediment cascade may also be driven by extrinsic change, such as climatic change or anthropogenic modification of the landscape (Chiverrell and Jakob 2013). Where timescales of extrinsic change are shorter than the timescales of intrinsic change in the sediment cascade, then approaches which assume stationarity in carbon flux through the sediment cascade will be flawed (Kuhn et al. 2009). Therefore, integration of the insights of climate modelling and of studies of long-term environmental change are required to understand what is an appropriate averaging time for carbon storage estimates at the landscape scale. The challenge of estimating changing carbon storage under changing environmental conditions is significant, but geomorphological understanding is integral to this research challenge because as noted above, under these conditions, the reworking of existing carbon stores by erosional processes has the potential to become the dominant control on net terrestrial carbon balance at the catchment scale.

The need for geomorphological understanding: As with the previous section, an important component of addressing this research challenge is the integration of existing geomorphological and geochronological data into our understanding of carbon flux. In many cases, this simply requires calibration of the carbon content of measured sediment fluxes. Integration of sediment budget approaches (see Chapter 4), with understanding of processes driving carbon transformations in the landscape, is required to appropriately analyse conditions of non-equilibrium carbon flux. Recalibration of sediment budget data as particulate carbon budgets through incorporation of measurements of organic matter content (e.g., Evans et al. 2006) provides a starting point for these investigations, but consideration of the appropriate timescale for sediment budget construction (see previous section) is essential.

A Method for the Analysis of Carbon Land Systems

The empirical evidence from a range of carbon landsystems along the sediment cascade described in Chapters 5–8 clearly demonstrates the importance of integrating geomorphological, biological and microbiological knowledge to understand the terrestrial carbon cycle. As discussed in Chapter 9, the research in this area underpins initiatives to modify terrestrial ecosystems to maximise carbon sequestration and storage. Integration of these fields of knowledge is fundamentally a multidisciplinary effort. In particular, the rapid technical advances in genetic characterisation of microbial communities and processes require domain specific expertise to fully exploit their potential. Integrating multi-disciplinary teams is most effective where there is a clear conceptual model of the problem being adressed, so that the contributions of specialists become more than the sum of the parts. In the previous sections of this chapter, a model of the key controls on the terrestrial carbon cycle that includes biological, microbiological and geomorphological controls has been explored. The stability of carbon sequestering landforms and their longevity in the landscape influences the relative importance of these three controls. In the following sections, a practical approach to operationalising investigation of the terrestrial carbon cycle through this conceptual model is outlined.

Beyond scientific curiosity, the requirement to understand the terrestrial carbon cycle and the sediment cascade is rooted in the potential to manage the large terrestrial carbon sink as part of mitigation of increasing atmospheric carbon concentrations. The approach outlined here defines three analytical steps which are required to characterise the landscape carbon system in order to plan effective interventions. These are: 1) full inventory and quantification of sedimentary carbon storage; 2) quantification of carbon fluxes, which is effectively a spatially distributed quantification of the terms in Equation 10.1; and 3) identification of fundamental processes and drivers of change in carbon–sediment systems. These three steps are explored in the following sections.

Quantifying Carbon Storage: From Sediment Budgets to Landscape Carbon Budgets

The challenges of describing material flux through sedimentary systems that respond to external and internal drivers of change at varying timescales has been approached in the geomorphological literature, through the application of sediment budgeting ideas. In a seminal paper, Dietrich et al. (1982) detail a practical approach to the challenges of defining catchment sediment budgets, as discussed in Chapter 4. The basic tenets of building a sediment budget, which include identification and quantification of sediment stores, and the processes which transfer sediment between stores, are a valuable starting point for outlining a geomorphologically informed method for the analysis of carbon landsystems.

Carbon storage within the sediment cascade is a function of the volume and carbon density of depositional landforms across a landscape. Storage of carbon in vegetation is not considered here, since at multi-annual timescales and large spatial scales (e.g., so that local forest disturbance is averaged across the landscape) the vegetation store is assumed to be in equilibrium. This is of course not the case where widespread transformation of vegetation cover types occurs, for example in response to climate change, in which case, closer attention to the vegetation store is required.

Traditional geomorphological classification of landforms and mapping approaches aid identification of these zones of storage (e.g., Swinnen et al. 2020) and modern geomorphology has taken advantage of advances in satellite and airborne remote sensing to support landscape scale mapping of deposits (Smith and Pain 2009). Field sampling of representative sites is required to assess carbon density and apply appropriate geospatial analysis to extrapolate to landscape scale estimates of carbon storage within the sediment cascade (e.g., Alderson et al. 2019a).

Quantifying Contemporary Carbon Flux

Equation 10.1 comprises three elements which are: 1) the net balance of NPP and decomposition; 2) the net balance of carbon erosion and deposition; and 3) the balance of DOC and DIC flux at a point in the system. A spatially distributed quantification of these terms is the basis for describing landscape carbon dynamics, and requires a multi-disciplinary range of methodological approaches as described below.

Quantification of NPP and Decomposition

Calculation of Net Ecosystem Exchange (NEE), which is the balance of primary productivity and soil respiration, is a major focus of research for biogeochemists. Measured in the field, this parameter represents the balance at a point in space between carbon uptake by plants through photosynthesis and release of carbon to the atmosphere through ecosystem respiration. Point measurements are often based on measurement of changing carbon dioxide concentration within chambers which isolate parts of the surface (e.g., Virkkala et al. 2018). Scaling up these measurements to time and space scales suitable for considering carbon budgets for the sediment cascade is challenging (Laine et al. 2006) and requires significant fieldwork time to develop sufficient datasets for extrapolation. Alternative approaches have the potential to measure carbon flux at larger spatial scales. For example, eddy covariance approaches (e.g., Baldocchi 2020) assess vertical fluxes of CO_2 and measure NEE over a footprint hundreds of metres upwind of the measurement point, and represent integration at spatial scales circa four orders of magnitude greater than chamber measurements.

At an even larger scale, satellite remote sensing approaches to estimate carbon flux, which combine observations of carbon dioxide concentrations with models of atmospheric redistribution (Crowell et al. 2018), are beginning to offer the potential to assess landscape scale carbon fluxes. Similarly, optical remote sensing approaches offer the potential to estimate NPP at the landscape scale (Šímová and Storch 2017; Song et al. 2013). In combination, these approaches offer the capacity to directly estimate carbon flux and NEE at scales compatible with the analysis of the sediment cascade. However, these approaches are still at the research frontier, and analysis at both the small scale (to allow integration with microbiological understanding of the processes of organic matter decomposition) and at larger scales compatible with analysis of the sediment cascade are required.

Quantification of Erosion and Deposition

This is a core area of geomorphological competence. A range of traditional geomorphological approaches are relevant, including direct measurement of erosion and deposition through approaches such as erosion pins and sediment traps and repeat surveys (e.g., Sirvent et al. 1997). These approaches are gradually being supplanted by the development of technologies which allow repeat collection of high resolution DEM data, providing spatially distributed erosion estimates. Techniques such as aerial and ground based LiDAR, Structure from Motion photography, and photogrammetry from drone surveys based on visual imaging, allow rapid generation of DEMs, which in turn allows estimation of net erosion rates at the landform to landscape scale (e.g., Glendell et al. 2017), at a range of timescales from event to multi-annual.

Assessment of integrated net erosion rates at the catchment scale is typically derived from measurement of fluvial sediment flux. Traditional approaches extrapolate point sediment concentration data, using long-term hydrometric monitoring using rating curve approaches (Asselman 2000). Continuous sediment flux data can also be derived from well calibrated continuous turbidity measurements (e.g., Wass and Leeks 1999) or from acoustic doppler data which has particular applicability in large river systems (e.g., Kostaschuk et al. 2005). In the largest river systems, satellite remote sensing of sediment flux calibrated by in stream measurement is also possible (e.g., Mangiarotti et al. 2013). Relevant timescales for these approaches to estimate net erosion of landforms and of wider catchment systems are on the order of days to decades, with the longest timescales relating to data from long established hydrometric networks and so limited in their spatial extent. Longer-term net erosion data can be derived from approaches which utilise tracers within the landscape to assess accumulation or erosion at a point in space. A widely applied approach is the use of caesium fallout records, to assess field scale net erosion rates subsequent to the bomb testing related caesium peak in the 1950s (Walling and Quine 1990). The approach allows decadal scale rates of erosion to be assessed across the landscape, but sampling densities are limited by the time and cost constraints of dating.

Quantification of Dissolved Carbon Fluxes

DOC and DIC flux through fluvial systems can be estimated from measurements of stream water concentrations and river discharge records. In locations with effective national hydrometric networks, these data can be combined with targeted measurement of water samples to extrapolate to annual fluxes (e.g., Worrall and Burt 2007). More recently, the potential to monitor DOC directly has begun to offer the opportunity to directly monitor flux in bespoke locations (Lee et al. 2015).

Identification of Fundamental Processes and Drivers of Change in Carbon-Sediment System

A complete understanding of carbon cycling in the sediment cascade requires not just the quantification of Equation 10.1 as described above, but also an understanding of the drivers of change in the carbon system, as described in Figures 10.1 and 10.2. A key finding of the empirical chapters of this book is that a process-based understanding of carbon cycling at the scale of depositional landforms within the sediment cascade requires better understanding of the microbial processes that drive the mineralisation of OM in sedimentary storage. As noted in the preceding section, recent research suggests that environmental constraints on microbial metabolism are the ultimate driver of carbon sequestration in sedimentary systems. Where microbial populations, OM production and environmental conditions are in equilibrium, then the details of microbial metabolism can be subsumed into environmental proxies, so that relations between, for example, local soil water saturation and rates of OM mineralisation are relatively constant. However, one of the key challenges of understanding terrestrial carbon cycling under conditions of climate change is the risk that the system moves into non-analogue conditions, where the nature of the known environmental response curves change.

As is apparent from Chapters 5–8, understanding of microbial drivers of change is an evolving area of research. Much previous research has focussed on characterising microbial communities, but as noted by Andersen et al. (2013) in relation to peatland systems (Chapter 5), closer attention to linking the structure of these communities with function is required in order to understand the environmental dependence of the microbial drivers of mineralisation of organic carbon. Advances in the technology for genetic characterisation of microbial community and function mean that the potential to address this research need is expanding rapidly. In particular, the potential of low cost 'lab on a chip' technologies (Liu and Zhu 2005) and the possibility to use RNA-based techniques to study gene activity (e.g., Hønsvall and Robertson 2017) enhance the capability to link microbial processes to environmental drivers at appropriate time and space scales. Linkage of microbial dynamics to gas flux at timescales of hours to weeks, and to longer-term change in vegetation and sedimentary conditions through the study of chronosequences (e.g., Allison et al. 2007), offers the potential to understand the interaction between short carbon dynamics driven by hydrometeorological change and longer-term change due to biological and geomorphological change. Higher spatial frequency analysis of microbial processes also offers the potential to understand small-scale variability in microbial niches within sediment systems, and to upscale from point measurements. The linkage of microbial function to carbon flux in sedimentary systems is very much at the research frontier, but the potential exists to develop process-based understanding that will ultimately support modelling of no analogue conditions.

Defining Appropriate Spatial Scales for the Analysis of the Carbon Sediment System

The methods available for the quantification of sediment and carbon fluxes described above range from direct measurement at the plot scale, to increasing application of remote sensing approaches to estimate carbon and sediment fluxes at the landscape scale. The emergence of high frequency remotely sensed data offer the potential to characterise carbon fluxes at large scales relevant to consideration of the whole sediment cascade. However, plot and landform level data are still essential to the development of process understanding, in particular linking microbial community function and dynamics to carbon release from OM in the sedimentary system. The ability to link this process understanding to measures of large-scale carbon flux requires an approach to the spatial extrapolation of point data.

The proposal by Marschner et al. (2008) that all OM is degradable at timescales of around 50 years suggests that it is not the inherent recalcitrance of some forms of OM which drive carbon sequestration, but environmental conditions which supress microbial decomposition either through limiting microbial metabolism, or through making OM inaccessible (e.g., through binding to mineral sediment). Under this hypothesis, the role of local environmental conditions becomes the key control on carbon flux, and therefore a useful basis for extrapolation of the fluxes which derive from the interaction of stored OM and processes of microbial decomposition.

Alderson et al. (2016) proposed that the dominant control on microbial metabolism of carbon from sedimentary stores was the aerobic status of the sediments. Anaerobic microbial respiration is associated with slower rates of OM decomposition. Consequently, a key control on the carbon preservation potential of sedimentary carbon stores is landscape position. As discussed above, propensity for saturation at a site can be estimated from digital elevation data using parameters such as the topographic wetness index which combines drainage potential (slope) and water supply (contributing area) (Beven and Kirkby 1979). This parameter therefore provides a useful sampling frame for studies of microbial activity and carbon decomposition rates at the landscape scale, with the potential to extrapolate findings from plot scale work to the whole landscape based on simple digital elevation data.

Remotely sensed data offers the potential to measure carbon fluxes at the spatial scale of the landscape unit, which is appropriate for understanding the role of the sediment cascade in terrestrial carbon cycling. However, in order to understand the process basis of these observations, the integration of higher resolution data is required. The ability to extrapolate these data to larger spatial scales, so that the estimates are validated by large-scale measurements, is a test of the degree to which the conceptual models and process understanding underpinning the extrapolation is an adequate description of the system.

Defining Appropriate Timescales for Analysis of the Carbon Sediment System

As discussed previously, one of the challenges of compiling the carbon balance for elements of the sediment cascade is the mismatch of timescales between the physical and biological elements of the system. Most of the measurement approaches described above allow characterisation of short-term change in the carbon balance, but the longer timescales over which change in the sediment system occurs mean that appropriate timescales for analysis may be longer than the days to decades discussed above. Distinguishing between equilibrium carbon dynamics (which may require consideration of century to millennial timescales) and short-term rates of change in disequilibrium conditions is important in the context of managing and intervening in the system.

The starting point for managed interventions in the terrestrial carbon cycle is the contemporary measured carbon balance as described above. The aim of interventions aimed at mitigating climate change is to shift the contemporary carbon balance to increase carbon sequestration or reduce carbon loss. These benefits may be permanent changes in the equilibrium carbon balance (e.g., through the restoration of eroding peatlands which become long-term carbon stores) or they may be temporary increases in carbon storage (e.g., through interventions which mimic disturbance and take advantage of early successional increases in NPP).

Understanding the longer-term impacts of interventions in the sediment–carbon system requires clarity on the nature of the long-term equilibrium conditions. For example, floodplains store significant amounts of sediment, and floodplain accretion is supported by sediment inputs from the upstream catchment system. The addition of pulses of sediment to a system (whether through the impacts of climate change or anthropogenic catchment change) leads to transmission of this pulse through the fluvial system, controlled by processes intrinsic to the system (Schumm 1973; Trimble 1981). On floodplains where the local sediment dynamics shift from depositional to erosional conditions, carbon loss controlled by these intrinsic thresholds can shift the floodplain to become a net carbon source, rather than a net sink of carbon. River management to preserve the carbon store is an immediate net gain; however, the longevity of that gain relates to the rate of net carbon loss from the eroding floodplain, since the carbon store is finite and eventually rates of carbon loss will become supply limited. Therefore, in order to distinguish between short-term gains (related to avoidance of the loss of stored carbon) from the establishment of a new equilibrium system state which sequesters carbon, the longer timescales of change in the sedimentary system become the appropriate level of analysis. The application of stratigraphic approaches with suitable geochronological control (e.g., Alderson et al. 2019a; Hoffman 2015) is therefore a necessary part of the characterisation of the carbon landsystem.

At the landscape scale, the optimum condition for a carbon sequestering landscape will be indicated by the identification of carbon stores with three key characteristics:

1) Sedimentary stores of carbon are large relative to the measured annual fluvial carbon flux and relative to measured rates of Net Ecosystem Exchange.
2) Carbon stores are persistent in the landscape. Application of geochronological approaches, particularly radiocarbon dating which directly ages the stored carbon, are important to establish timescales of carbon storage. Long-term

stability of the carbon stores are indicative of landscapes where the sediment system is closer to dynamic equilibrium

3) Carbon stores are accumulating carbon. This means that the carbon flux, as expressed by Equation 10.1, indicates contemporary carbon accumulation.

These are characteristics of the carbon landsystem and are influenced by both biological and geomorphological processes.

Conclusions

Carbon dynamics in the sediment cascade are controlled by sediment dynamics operating at timescales of centuries to millennia, vegetation dynamics operating at timescales of decades to centuries, and microbial dynamics operating at time-scales of days to weeks. Where intrinsic or extrinsic change in the sediment system drives physical instability in sediment stores, then these long-term geomorphological processes are likely to dominate the carbon balance. Where there is physical stability of carbon sequestering landforms, the interaction of vegetation and microbial dynamics control the addition of autochthonous carbon to the sediment store. Sedimentary carbon storage in depositional landforms represents a balance of allochthonous delivery of carbon through geomorphological processes and the sequestration of autochthonous carbon sources controlled by biological and microbiological processes.

Fluxes of terrestrial carbon from headwaters to the ocean are represented as a black box in most carbon models. Cole et al. (2007) called for a focus on 'plumbing' the carbon cycle in order to understand the nature of fluxes through the fluvial system. What this book has tried to demonstrate is that it is not only the plumbing that needs attention, but the fabric of the fluvial system which the plumbing connects. Contemporary concern with the terrestrial carbon cycle is rooted in the effects of elevated carbon concentrations in the atmosphere. Therefore, the significance of changes in sediment associated carbon fluxes rest critically in whether that carbon is mineralised and released to the atmosphere, or whether it is simply transferred to another depositional store. Determining the fate of carbon in sedimentary storage requires data on sediment quality, on microbial processes which drive mineralisation and on response of these microbial systems to changing environmental conditions.

As this book has demonstrated, multidisciplinary approaches to the characterisation of these processes require an explicit focus on spatial and temporal scales in order to distinguish transient from long-term change, and equilibrium conditions from conditions of local instability. Technological change holds the prospect of rapid progress in these areas over the next decade. The rapid development of remote sensing approaches to the analysis of geomorphic change, vegetation change and large-scale gas flux offer the potential to integrate site specific understanding at the landscape scale. Similarly, the exponential changes in capacity to sequence genetic material and the development of 'lab on a chip' approaches offer the capacity to move beyond the description of microbial systems, to understand functional change at larger spatial scales and shorter timescales than has previously been possible. Critically however, developments in understanding carbon flux through the sediment cascade at larger spatial scales require consideration of the longer timescales at which geomorphological change in that cascade can occur. This requires the insights not just of process geomorphology, but of the understanding of landform history and evolution which derives from stratigraphic and chronological approaches to analysis of the landscape system. The increasing focus on the use of radiocarbon analysis to understand the age of the carbon being cycled through the system is an important development in support of this work.

This book argues that analysis of the terrestrial carbon cycle, which does not engage with geomorphological understanding, is incomplete. At the same time, a focus on carbon storage and transport through the sediment cascade is a partial analysis if it does not consider the processes which drive carbon sequestration or release from sedimentary storage. Progress in this area requires a diverse range of specialist expertise and requires multidisciplinary teams encompassing biologists, microbiologists, biogeochemists, geomorphologists, remote sensors and modellers. The preceding chapters have reviewed the state of the science relating to understanding carbon flux through different terrestrial sediment systems. Despite the considerable body of research in this area, the ability to develop an integrated understanding is limited by the paucity of multi-disciplinary investigations. It is clear that further empirical data that measures rates of carbon transformation in the vegetation system, the soil microbiological system and the geomorphological systems, and the interdependencies between these elements for the major storage elements in the carbon land system, is required. This final chapter has proposed a framework to take this work forward.

As has been described, much of the expertise required to do this work exists within existing disciplinary boundaries, and beyond this the technological capacity to carry out this science and to support the necessary integration of scales of analysis is evolving rapidly. The next decade offers an exciting prospect for developments in this field. Given the rapidly escalating concern over human impacts on carbon cycling, and the potential to mitigate these effects, this research will be more than timely.

References

Aagaard, T. and Sørensen, P. 2013. Sea level rise and the sediment budget of an eroding barrier on the Danish North Sea coast. *Journal of Coastal Research* 65, 434–439. doi: 10.2112/SI65-074.1

Aalto, R., Lauer, J.W., and Dietrich, W.E. 2008. Spatial and temporal dynamics of sediment accumulation and exchange along Strickland River floodplains (Papua New Guinea) over decadal-to-centennial timescales. *Journal of Geophysical Research Earth Surface* 113, F1, 1–22. doi: 10.1029/2006JF000627

Abramoff, R., Xu, X., Hartman, M. et al. 2018. The Millennial model: In search of measurable pools and transformations for modeling soil carbon in the new century. *Biogeochemistry* 137, 51–71. doi: 10.1007/s10533-017-0409-7

Abril, G., Nogueira, M., Etcheber, H. et al. 2002. Behaviour of organic carbon in nine contrasting European estuaries. *Estuarine, Coastal and Shelf Science* 54, 2, 241–262. doi: 10.1006/ecss.2001.0844

Adame, M.F., Neil, D., Wright, S.F. et al. 2010. Sedimentation within and among mangrove forests along a gradient of geomorphological settings. *Estuarine, Coastal and Shelf Science* 86, 1, 21–30. doi: 10.1016/j.ecss.2009.10.013

Aitkenhead, J.A. and McDowell, W.H. 2000. Soil C:N ratio as a predictor of annual riverine DOC flux at local and global scales. *Global Biogeochemical Cycles* 14, 1, 127–138. doi: 10.1029/1999GB900083

Alderson, D.M., Evans, M.G., Rothwell, J.J. et al. 2016. Classifying sedimentary organics: It is a matter of quality rather than quantity. *Progress in Physical Geography* 40, 3, 450–479. doi: 10.1177/0309133315625864

Alderson, D.M., Evans, M.G., Rothwell, J.J. et al. 2019a. Geomorphological controls on fluvial carbon storage in headwater peatlands. *Earth Surface Processes and Landforms* 44, 9, 1675–1693. doi: 10.1002/esp.4602

Geomorphology and the Carbon Cycle, First Edition. Martin Evans.

© 2022 Royal Geographical Society (with the Institute of British Geographers). Published 2022 by John Wiley & Sons Ltd.

Alderson, D.M., Evans, M.G., and Shuttleworth, E.L. 2019b. Trajectories of ecosystem change in restored blanket peatlands. *Science of the Total Environment* 665, 785–796. doi: 10.1016/j.scitotenv.2019.02.095

Alin, S.R., Aalto, R., Goni, M.A. et al. 2008. Biogeochemical characterization of carbon sources in the Strickland and Fly rivers, Papua New Guinea. *Journal of Geophysical Research Earth Surface* 113, F1, 1–21. doi: 10.1029/2006JF000625

Allen, J. 2000. Morphodynamics of Holocene salt marshes: A review sketch from the Atlantic and Southern North Sea coasts of Europe. *Quaternary Science Reviews* 19, 12, 1155–1231. doi: 10.1016/S0277-3791(99)00034-7

Allison, V.J., Condron, L.M., Peltzer, D.A. et al. 2007. Changes in enzyme activities and soil microbial community composition along carbon and nutrient gradients at the Franz Josef chronosequence, New Zealand. *Soil Biology & Biochemistry* 39, 7, 1770–1781. doi: 10.1016/j.soilbio.2007.02.006

Alongi, D.M. 2014. Carbon cycling and storage in mangrove forests. *Annual Review of Marine Science* 6, 195–219. doi: 10.1146/annurev-marine-010213-135020

Alongi, D.M., Pfitzner, J., Trott, L.A. et al. 2005. Rapid sediment accumulation and microbial mineralization in forests of the mangrove Kandelia candel in the Jiulongjiang Estuary, China. *Estuarine, Coastal and Shelf Science* 63, 4, 605–618. doi: 10.1016/j.ecss.2005.01.004

Andersen, R., Chapman, S.J., and Artz, R.R.E. 2013. Microbial communities in natural and disturbed peatlands: A review. *Soil Biology and Biochemistry* 57, 979–994. doi: 10.1016/j.soilbio.2012.10.003

Anderson, M.G. and Burt, T.P. 1978. The role of topography in controlling through-flow generation. *Earth Surface Processes and Landforms* 3, 4, 331–344. doi: 10.1002/esp.3290030402

Anderson, P., Buckler, M., and Walker, J. 2009. Moorland restoration: Potential and progress, in Bonn, A., Allott, T.E.H., Evans, M. et al. (eds.), *Peatland Restoration and Ecosystem Services: Science Policy and Practice.* Cambridge University Press, Cambridge, UK, pp. 432–447.

Anderson, S.P. 2005. Glaciers show direct linkage between erosion rate and chemical weathering fluxes. *Geomorphology* 67, 1–2, 147–157. doi: 10.1016/j.geomorph.2004.07.010

Anderson, S.P. 2007. Biogeochemistry of glacial landscape systems. *Annual Review of Earth and Planetary Sciences* 35, 1, 375–399. doi: 10.1146/annurev.earth.35.031306.140033

Anesio, A.M., Hodson, A.J., Fritz, A. et al. 2009. High microbial activity on glaciers: Importance to the global carbon cycle. *Global Change Biology* 15, 4, 955–960. doi: 10.1111/j.1365-2486.2008.01758.x

Anesio, A.M., Sattler, B., Foreman, C. et al. 2010. Carbon fluxes through bacterial communities on glacier surfaces. *Annals of Glaciology* 51, 56, 32–40. doi: 10.3189/172756411795932092

Anselmetti, F.S., Hodell, D.A., Ariztegui, D. et al. 2007. Quantification of soil erosion rates related to ancient Maya deforestation. *Geology* 35, 10, 915–918. doi: 10.1130/G23834A.1

Archer, D. 2008. Checking the thermostat. *Nature Geoscience* 1, 289–290. doi: 10.1038/ngeo194

Aronson, J. and Alexander, S. 2013. Ecosystem restoration is now a global priority: Time to roll up our sleeves. *Restoration Ecology* 21, 3, 293–296. doi: 10.1111/rec.12011

Arunachalam, A. and Upadhyaya, K. 2005. Microbial biomass during revegetation of landslides in the humid tropics. *Journal of Tropical Forest Science* 17, 2, 306–311. https://www.jstor.org/stable/23616577

Aselmann, I. and Crutzen, P.J. 1989. Global distribution of natural freshwater wetlands and rice paddies, their net primary productivity, seasonality and possible methane emissions. *Journal of Atmospheric Chemistry* 8, 4, 307–358. doi: 10.1007/BF00052709

Ashmore, P. and Dodson, B. 2017. Urbanizing physical geography. *The Canadian Geographer* 61, 1, 102–106. doi: 10.1111/cag.12318

Asselman, N.E.M. 2000. Fitting and interpretation of sediment rating curves. *Journal of Hydrology* 234, 3–4, 228–248. doi: 10.1016/S0022-1694(00)00253-5

Aufdenkampe, A.K., Mayorga, E., Raymond, P.A. et al. 2011. Riverine coupling of biogeochemical cycles between land, oceans, and atmosphere. *Frontiers in Ecology and the Environment* 9, 1, 53–60. doi: 10.1890/100014

Baguette, M., Blanchet, S., Legrand, D. et al. 2013. Individual dispersal, landscape connectivity and ecological networks. *Biological Reviews* 88, 2, 310–326. doi: 10.1111/brv.12000

Baker, B.J., Lazar, C.S., Teske, A.P. et al. 2015. Genomic resolution of linkages in carbon, nitrogen, and sulfur cycling among widespread estuary sediment bacteria. *Microbiome* 3, 14, 1–12. doi: 10.1186/s40168-015-0077-6

Baldocchi, D.D. 2020. How eddy covariance flux measurements have contributed to our understanding of Global Change Biology. *Global Change Biology* 26, 1, 242–260. doi: 10.1111/gcb.14807

Baldwin, D.S., Colloff, M.J., Mitrovic, S.M. et al. 2016. Restoring dissolved organic carbon subsidies from floodplains to lowland river food webs: A role for environmental flows? *Marine and Freshwater Research* 67, 9, 1387–1399. doi: 10.1071/MF15382

Balke, T. and Friess, D.A. 2016. Geomorphic knowledge for mangrove restoration: A pan-tropical categorization. *Earth Surface Processes and Landforms* 41, 2, 231–239. doi: 10.1002/esp.3841

Bardgett, R.D., Richter, A., Bol, R. et al. 2007. Heterotrophic microbial communities use ancient carbon following glacial retreat. *Biology Letters* 3, 5, 487–490. doi: 10.1098/rsbl.2007.0242

Barker, J.D., Sharp, M.J., Fitzsimons, S.J. et al. 2006. Abundance and dynamics of dissolved organic carbon in glacier systems. *Arctic, Antarctic and Alpine Research* 38, 2, 163–172. doi: 10.1657/1523-0430(2006)38[163:AADODO]2.0.CO;2

Battin, T.J., Kaplan, L.A., Findlay, S. et al. 2008. Biophysical controls on organic carbon fluxes in fluvial networks. *Nature Geoscience* 1, 95–100. doi: 10.1038/ngeo101

Battin, T.J., Luyssaert, S., Kaplan, L.A. et al. 2009. The boundless carbon cycle. *Nature Geoscience* 2, 598–600. doi: 10.1038/ngeo618

Bauer, J.E., Cai, W.-J., Raymond, P.A. et al. 2013. The changing carbon cycle of the coastal ocean. *Nature* 504, 61–70. doi: 10.1038/nature12857

Beaumont, N.J., Jones, L., Garbutt, A. et al. 2014. The value of carbon sequestration and storage in coastal habitats. *Estuarine, Coastal and Shelf Science* 137, 32–40. doi: 10.1016/j.ecss.2013.11.022

Bechtold, J.S. and Naiman, R.J. 2009. A quantitative model of soil organic matter accumulation during floodplain primary succession. *Ecosystems* 12, 8, 1352–1368. doi: 10.1007/s10021-009-9294-9

Beckman, N.D. and Wohl, E. 2014. Carbon storage in mountainous headwater streams: The role of old-growth forest and logjams. *Water Resources Research* 50, 3, 2376–2393. doi: 10.1002/2013WR014167

Bednarek, A.T. 2001. Undamming rivers: A review of the ecological impacts of dam removal. *Environmental Management* 27, 803–814. doi: 10.1007/s002670010189

Beer, C., Reichstein, M., Tomelleri, E. et al. 2010. Terrestrial gross carbon dioxide uptake: Global distribution and covariation with climate. *Science* 329, 5993, 834–838. doi: 10.1126/science.1184984

Benda, L., Hassan, M.A., Church, M. et al. 2005. Geomorphology of steepland headwaters: The transition from hillslopes to channels. *Journal of the American Water Resources Association* 41, 4, 835–851. https://doi.org/10.1111/j.1752-1688.2005.tb03773.x

Benda, L., Poff, N.L., Miller, D. et al. 2004. The network dynamics hypothesis: How channel networks structure riverine habitats. *BioScience* 54, 5, 413–427. doi: 10.1641/0006-3568(2004)054[0413:TNDHHC]2.0.CO;2

Benn, D. and Evans, D.J.A. 2014. *Glaciers and Glaciation*. Routledge, Abingdon, UK.

Benner, R., Newell, S.Y., Maccubbin, A.E. et al. 1984. Relative contributions of bacteria and fungi to rates of degradation of lignocellulosic detritus in salt-marsh sediments. *Applied and Environmental Microbiology* 48, 1, 36–40. https://doi.org/10.1128/aem.48.1.36-40.1984

Berhe, A.A. 2012. Decomposition of organic substrates at eroding vs. depositional landform positions. *Plant and Soil* 350, 1–2, 261–280. doi: 10.1007/s11104-011-0902-z

Berhe, A.A., Barnes, R.T., Six, J. et al. 2018. Role of soil erosion in biogeochemical cycling of essential elements: Carbon, nitrogen, and phosphorus. *Annual Review of Earth and Planetary Sciences* 46, 1, 521–548. doi: 10.1146/annurev-earth-082517-010018

Berhe, A.A., Harden, J.W., Torn, M.S. et al. 2012. Persistence of soil organic matter in eroding versus depositional landform positions. *Journal of Geophysical Research: Biogeosciences* 117, G2, 1–16. doi: 10.1029/2011JG001790

Berhe, A.A., Harte, J., Harden, J.W. et al. 2007. The significance of the erosion-induced terrestrial carbon sink. *BioScience* 57, 4, 337–346. doi: 10.1641/B570408

Berhe, A.A. and Kleber, M. 2013. Erosion, deposition, and the persistence of soil organic matter: Mechanistic considerations and problems with terminology. *Earth Surface Processes and Landforms* 38, 8, 908–912. doi: 10.1002/esp.3408

Bernal, B. and Mitsch, W.J. 2012. Comparing carbon sequestration in temperate freshwater wetland communities. *Global Change Biology* 18, 5, 1636–1647. doi: 10.1111/j.1365-2486.2011.02619.x

Berner, A. and Caldeira, K. 1997. The need for mass balance and feedback in the geochemical carbon cycle. *Geology* 25, 10, 955–956. doi: 10.1130/0091-7613(1997)025<0955:TNFMBA>2.3.CO;2

Berner, R.A. 1997. The rise of plants and their effect on weathering and atmospheric CO_2. *Science* 276, 5312, 544–546. doi: 10.1126/science.276.5312.544

Berner, R.A. 2003. The long-term carbon cycle, fossil fuels and atmospheric composition. *Nature* 426, 323–326. doi: 10.1038/nature02131

Bernhardt, E.S., Palmer, M.A., Allan, J.D. et al. 2005. Synthesizing U.S. river restoration efforts. *Science* 308, 5722, 636–637. doi: 10.1126/science.1109769

Besemer, K., Singer, G., Quince, C. et al. 2013. Headwaters are critical reservoirs of microbial diversity for fluvial networks. *Proceedings of the Royal Society B: Biological Sciences* 280, 1771, 1–8. 20131760–20131760. doi: 10.1098/rspb.2013.1760

Beven, K.J. and Kirkby, M.J. 1979. A physically based, variable contributing area model of basin hydrology. *Hydrological Sciences Bulletin* 24, 1, 43–69. doi: 10.1080/02626667909491834

Beyer, L., Blume, H.-P., Elsner, D.-C. et al. 1995. Soil organic matter composition and microbial activity in urban soils. *Science of the Total Environment* 168, 3, 267–278. doi: 10.1016/0048-9697(95)04704-5

Billett, M.F., Charman, D.J., Clark, J.M. et al. 2010. Carbon balance of UK peatlands: Current state of knowledge and future research challenges. *Climate Research* 45, 13–29. doi: 10.3354/cr00903

Bishop, K., Buffam, I., Erlandsson, M. et al. 2008. Aqua incognita: The unknown headwaters. *Hydrological Processes* 22, 8, 1239–1242. doi: 10.1002/hyp.7049

Blair, N.E. and Aller, R.C. 2012. The fate of terrestrial organic carbon in the marine environment. *Annual Reviews* 4, 401–423. doi: 10.1146/annurev-marine-120709-142717

Bloom, A.A., Exbrayat, J.-F., van der Velde, I.R. et al. 2016. The decadal state of the terrestrial carbon cycle: Global retrievals of terrestrial carbon allocation, pools, and residence times. *Proceedings of the National Academy of Sciences of the United States of America* 113, 5, 1285–1290. doi: 10.1073/pnas.1515160113

Blum, L.K., Mills, A.L., Zieman, J.C. et al. 1988. Abundance of bacteria and fungi in seagrass and mangrove detritus. *Marine Ecology- Progress Series* 42, 73–78. doi: 10.3354/meps042073

Boardman, J. 2013. Soil erosion in Britain: Updating the record. *Agriculture* 3, 3, 418–442. doi: 10.3390/agriculture3030418

Bond-Lamberty, B., Bailey, V.L., Chen, M. et al. 2018. Globally rising soil heterotrophic respiration over recent decades. *Nature* 560, 7716, 80–83. doi: 10.1038/s41586-018-0358-x

Bond-Lamberty, B., Wang, C., and Gower, S.T. 2004. A global relationship between the heterotrophic and autotrophic components of soil respiration? *Global Change Biology* 10, 10, 1756–1766. doi: 10.1111/j.1365-2486.2004.00816.x

Bonn, A., Allott, T., Evans, M. et al. 2016. *Peatland Restoration and Ecosystem Services: Science, Policy and Practice.* Cambridge University Press, Cambridge, UK.

Borge, A.F., Westermann, S., Solheim, I. et al. 2017. Strong degradation of palsas and peat plateaus in northern Norway during the last 60 years. *The Cryosphere Discussions* 11, 1–16. doi: 10.5194/tc-11-1-2017

Borges, A.V., Delille, B., and Frankignoulle, M. 2005. Budgeting sinks and sources of CO_2 in the coastal ocean: Diversity of ecosystems counts. *Geophysical Research Letters* 32, 14, 1–4. doi: 10.1029/2005GL023053

Bouchez, J., Gaillardet, J., Lupker, M. et al. 2012. Floodplains of large rivers: Weathering reactors or simple silos? *Chemical Geology* 332–333, 166–184. doi: 10.1016/j.chemgeo.2012.09.032

Bouillon, S., Borges, A.V., Castañeda-Moya, E. et al. 2008. Mangrove production and carbon sinks: A revision of global budget estimates. *Global Biogeochemical Cycles* 22, 2, 1–12. doi: 10.1029/2007GB003052

Bouillon, S., Moens, T., Koedam, N. et al. 2004. Variability in the origin of carbon substrates for bacterial communities in mangrove sediments. *FEMS Microbiology Ecology* 49, 2, 171–179. doi: 10.1016/j.femsec.2004.03.004

Bower, M.M. 1961. The distribution of erosion in blanket peat bogs in the Pennines. *Transactions of the Institute of British Geographers* 29, 17–30. doi: 10.2307/621241

Bracken, L.J. and Croke, J. 2007. The concept of hydrological connectivity and its contribution to understanding runoff-dominated geomorphic systems. *Hydrological Processes* 21, 13, 1749–1763. doi: 10.1002/hyp.6313

Bracken, L.J., Turnbull, L., Wainwright, J. et al. 2015. Sediment connectivity: A framework for understanding sediment transfer at multiple scales. *Earth Surface Processes and Landforms* 40, 2, 177–188. doi: 10.1002/esp.3635

Bragazza, L., Parisod, J., Buttler, A. et al. 2013. Biogeochemical plant-soil microbe feedback in response to climate warming in peatlands. *Nature Climate Change* 3, 273–277. doi: 10.1038/nclimate1781

Brantley, H.L., Hagler, G.S.W., Deshmukh, P.J. et al. 2014. Field assessment of the effects of roadside vegetation on near-road black carbon and particulate matter. *Science of the Total Environment* 468–469, 120–129. doi: 10.1016/j.scitotenv.2013.08.001

Brantley, S.L. and Lebedeva, M. 2011. Learning to read the chemistry of regolith to understand the critical zone. *Annual Review of Earth and Planetary Sciences* 39, 387–416. https://doi.org/10.1146/annurev-earth-040809-152321

Breithaupt, J.L., Smoak, J.M., Smith, T.J. et al. 2012. Organic carbon burial rates in mangrove sediments: Strengthening the global budget. *Global Biogeochemical Cycles* 26, 3, 1–11. doi: 10.1029/2012GB004375

Bridges, J.S. 2009. *Rivers and Floodplains: Forms, Processes and Sedimentary Record*. Blackwell Science, Oxford, UK.

Brierley, G., Fryirs, K., and Jain, V. 2006. Landscape connectivity: The geographic basis of geomorphic applications. *Area* 38, 2, 165–174. doi: 10.1111/j.1475-4762.2006.00671.x

Brinson, M.M., Bradshaw, H.D., Holmes, R.N. et al. 1980. Litterfall, stemflow, and throughfall nutrient fluxes in an alluvial swamp forest. *Ecology* 61, 4, 827–835. doi: 10.2307/1936753

Brooks, S. and Stoneman, R. 1997. *Conserving Bogs – The Management Handbook*. The Stationery Office, Edinburgh, UK.

Brown, A., Toms, P., Carey, C. et al. 2013. Geomorphology of the Anthropocene: Time-transgressive discontinuities of human-induced alluviation. *Anthropocene* 1, 3–13. doi: 10.1016/j.ancene.2013.06.002

Brown, A.G., Carey, C., Erkens, G. et al. 2009. From sedimentary records to sediment budgets: Multiple approaches to catchment sediment flux. *Geomorphology* 108, 1–2, 35–47. doi: 10.1016/j.geomorph.2008.01.021

Brown, S., Miltner, E., and Cogger, C. 2012. Carbon sequestration potential in urban soils, in Lal, R. and Augustin, B. (eds.), *Carbon Sequestration in Urban Ecosystems*. Springer, Dordrecht, Netherlands, pp. 173–196. doi: 10.1007/978-94-007-2366-5_9

Brown, S.L., Goulsbra, C.S., and Evans, M.G. 2019. Controls on fluvial carbon efflux from eroding peatland catchments. *Hydrological Processes* 33, 3, 361–371. doi: 10.1002/hyp.13329

Brown, T.G. and Sear, D.A. 2008. "Natural" streams in Europe: Their form and role in carbon sequestration. AGU Fall Meeting Abstracts 51.

Brunsden, D. and Thornes, J.B. 1979. Landscape sensitivity and change. *Transactions of the Institute of British Geographers* 4, 4, 463–484. doi: 10.2307/622210

Buchan, A., Newell, S.Y., Butler, M. et al. 2003. Dynamics of bacterial and fungal communities on decaying salt marsh grass. *Applied and Environmental Microbiology* 69, 11, 6676–6687. doi: 10.1128/AEM.69.11.6676-6687.2003

Buitenhuis, E.T., Hashioka, T., and Quéré, C.L. 2013. Combined constraints on global ocean primary production using observations and models. *Global Biogeochemical Cycles* 27, 3, 847–858. doi: 10.1002/gbc.20074

Burchsted, D., Daniels, M., Thorson, R. et al. 2010. The river discontinuum: Applying beaver modifications to baseline conditions for restoration of forested headwaters. *BioScience* 60, 11, 908–922. doi: 10.1525/bio.2010.60.11.7

Burchsted, D. and Daniels, M.D. 2014. Classification of the alterations of beaver dams to headwater streams in northeastern Connecticut, U.S.A. *Geomorphology* 205, 36–50. doi: 10.1016/j.geomorph.2012.12.029

Burns, A. and Ryder, D.S. 2001. Response of bacterial extracellular enzymes to inundation of floodplain sediments. *Freshwater Biology* 46, 10, 1299–1307. doi: 10.1046/j.1365-2427.2001.00750.x

Burns, R.G., DeForest, J.L., Marxsen, J. et al. 2013. Soil enzymes in a changing environment: Current knowledge and future directions. *Soil Biology and Biochemistry* 58, 216–234. doi: 10.1016/j.soilbio.2012.11.009

Burns, R.G. and Dick, R.P. 2002. *Enzymes in the Environment: Activity, Ecology, and Applications*. CRC Press, New York.

Burt, T. and Allison, R.J. 2010. *Sediment Cascades: An Integrated Approach*. Wiley Blackwell, Oxford, UK.

Burt, T.P. and Butcher, D.P. 1985. Topographic controls of soil moisture distributions. *Journal of Soil Science* 36, 3, 469–486. doi: 10.1111/j.1365-2389.1985.tb00351.x

Burt, T.P. and Pinay, G. 2005. Linking hydrology and biogeochemistry in complex landscapes. *Progress in Physical Geography: Earth and Environment* 29, 3, 297–316. doi: 10.1191/0309133305pp450ra

Butman, D., Stackpoole, S., Stets, E. et al. 2016. Aquatic carbon cycling in the conterminous United States and implications for terrestrial carbon accounting. *Proceedings of the National Academy of Sciences of the United States of America* 113, 58–63. doi: 10.1073/pnas.1512651112

Butman, D.E., Wilson, H.F., Barnes, R.T. et al. 2015. Increased mobilization of aged carbon to rivers by human disturbance. *Nature Geoscience* 8, 112–116. doi: 10.1038/ngeo2322

Cabezas, A. and Comín, F.A. 2010. Carbon and nitrogen accretion in the topsoil of the Middle Ebro River Floodplains (NE Spain): Implications for their ecological restoration. *Ecological Engineering* 36, 5, 640–652. doi: 10.1016/j.ecoleng.2008.07.021

Cabezas, A., Comín, F.A., and Walling, D.E. 2009. Changing patterns of organic carbon and nitrogen accretion on the middle Ebro floodplain (NE Spain). *Ecological Engineering* 35, 10, 1547–1558. doi: 10.1016/j.ecoleng.2009.07.006

Cai, W.-J. 2011. Estuarine and coastal ocean carbon paradox: CO_2 sinks or sites of terrestrial carbon incineration? *Annual Review of Marine Science* 3, 123–145. doi: 10.1146/annurev-marine-120709-142723

Campbell, D.R., Lavoie, C., and Rochefort, L. 2002. Wind erosion and surface stability in abandoned milled peatlands. *Canadian Journal of Soil Science* 82, 1, 85–95. doi: 10.4141/S00-089

Carey, J.C., Tang, J., Templer, P.H. et al. 2016. Temperature response of soil respiration largely unaltered with experimental warming. *Proceedings of the National Academy of Sciences of the United States of America* 113, 48, 13797–13802. doi: 10.1073/pnas.1605365113

Carling, P.A., Glaister, M.S., and Flintham, T.P. 1997. The erodibility of upland soils and the design of preafforestation drainage networks in the United Kingdom. *Hydrological Processes* 11, 15, 1963–1980. doi: 10.1002/(SICI)1099-1085(199712)11:15<1963::AID-HYP542>3.0.CO;2-M

Carvalhais, N., Forkel, M., Khomik, M. et al. 2014. Global covariation of carbon turnover times with climate in terrestrial ecosystems. *Nature* 514, 213–217. doi: 10.1038/nature13731

Cavalli, M., Trevisani, S., Comiti, F. et al. 2013. Geomorphometric assessment of spatial sediment connectivity in small alpine catchments. *Geomorphology* 188, 31–41. doi: 10.1016/j.geomorph.2012.05.007

Caves, J.K., Jost, A.B., Lau, K.V. et al. 2016. Cenozoic carbon cycle imbalances and a variable weathering feedback. *Earth and Planetary Science Letters* 450, 152–163. doi: 10.1016/j.epsl.2016.06.035

Chaopricha, N.T. and Marín-Spiotta, E. 2014. Soil burial contributes to deep soil organic carbon storage. *Soil Biology and Biochemistry* 69, 251–264. doi: 10.1016/j.soilbio.2013.11.011

Charman, D. 2002. *Peatlands and Environmental Change*. John Wiley & Sons, Chichester, UK.

Chartin, C., Bourennane, H., Salvador-Blanes, S. et al. 2011. Classification and mapping of anthropogenic landforms on cultivated hillslopes using DEMs and soil thickness data – Example from the SW Parisian Basin, France. *Geomorphology* 135, 1–2, 8–20. doi: 10.1016/j.geomorph.2011.07.020

Chen, G., Chen, B., Yu, D. et al. 2016. Soil greenhouse gas emissions reduce the contribution of mangrove plants to the atmospheric cooling effect. *Environmental Research Letters* 11, 12, 1–10. doi: 10.1088/1748-9326/11/12/124019

Chen, S., Day, S.D., Wick, A.F. et al. 2013b. Mean residence time of global topsoil organic carbon depends on temperature, precipitation and soil nitrogen. *Global and Planetary Change* 100, 99–108. doi: 10.1016/j.gloplacha.2012.10.006

Chen, X., Wei, X., and Scherer, R. 2005. Influence of wildfire and harvest on biomass, carbon pool, and decomposition of large woody debris in forested streams of southern interior British Columbia. *Forest Ecology and Management* 208, 1–3, 101–114. doi: 10.1016/j.foreco.2004.11.018

Chen, Y., Day, S.D., Wick, A.F. et al. 2013a. Changes in soil carbon pools and microbial biomass from urban land development and subsequent post-development soil rehabilitation. *Soil Biology and Biochemistry* 66, 38–44. doi: 10.1016/j.soilbio.2013.06.022

Chimner, R.A. and Karberg, J.M. 2008. Long-term carbon accumulation in two tropical mountain peatlands, Andes Mountains, Ecuador. *Mires and Peat* 3, 1–10. http://www.mires-and-peat.net/pages/volumes/map03/map0304.php

Chiverrell, R. and Jakob, M. 2013. Radiocarbon dating: Alluvial fan/debris cone evolution and hazards, in Schneuwly-Bollschweiler, M., Stoffel, M., and Rudolf-Miklau. F. (eds.), *Dating Torrential Processes on Fans and Cones. Advances in Global Change Research.* Springer, Dordrecht, Netherlands, pp. 265–282. doi: 10.1007/978-94-007-4336-6_17

Chiverrell, R.C., Oldfield, F., Appleby, P.G. et al. 2008. Evidence for changes in Holocene sediment flux in Semer Water and Raydale, North Yorkshire, UK. *Geomorphology* 100, 1–2, 70–82. doi: 10.1016/j.geomorph.2007.04.035

Chorley, R.J. and Kennedy, B.A. 1971. *Physical Geography: A Systems Approach.* Prentice Hall, London.

Christian, C.S. and Stewart, G.A. 1947. North Australian regional survey 1946, Katherine-Darwin Region. Mimeo, C.S.I.R.O., Melbourne.

Church, M. 2005. Continental drift. *Earth Surface Processes and Landforms* 30, 1, 129–130. doi: 10.1002/esp.1183

Church, M. 2010. The trajectory of geomorphology. *Progress in Physical Geography: Earth and Environment* 34, 3, 265–286. doi: 10.1177/0309133310363992

Church, M., Kellerhals, R., and Day, T.J. 1989. Regional clastic sediment yield in British Columbia. *Canadian Journal of Earth Sciences* 26, 1, 31–45. doi: 10.1139/e89-004

Church, M. and Ryder, J.M. 1972. Paraglacial sedimentation: A consideration of fluvial processes conditioned by glaciation. *Geological Society of America Bulletin* 83, 10, 3059–3072. doi: 10.1130/0016-7606(1972)83[3059:PSACOF]2.0.CO;2

Church, M. and Slaymaker, O. 1989. Disequilibrium of Holocene sediment yield in glaciated British Columbia. *Nature* 337, 452–454. doi: 10.1038/337452a0

Ciais, P., Sabine, C., Bala, G. et al. 2013. Carbon and other biogeochemical cycles, in Stocker, T.F., Qin, D., Plattner, G.-K. et al. (eds.), *Climate Change 2013: The Physical Science Basis. Contribution of Working Group I to the Fifth Assessment Report of the Intergovernmental Panel on Climate Change.* Cambridge University Press, Cambridge, UK and New York, pp. 465–570. doi: 10.1017/CBO9781107415324.01

Clair, T.A., Pollock, T.L., and Ehrman, J.M. 1994. Exports of carbon and nitrogen from river basins in Canada's Atlantic Provinces. *Global Biogeochemical Cycles* 8, 4, 441–450. doi: 10.1029/94GB02311

Clark, K.E., West, A.J., Hilton, R.G. et al. 2016. Storm-triggered landslides in the Peruvian Andes and implications for topography, carbon cycles, and biodiversity. *Earth Surface Dynamics* 4, 1, 47–70. doi: 10.5194/esurf-4-47-2016

Colangelo, D.J. 2007. Response of river metabolism to restoration of flow in the Kissimmee River, Florida, USA. *Freshwater Biology* 52, 3, 459–470. doi: 10.1111/j.1365-2427.2006.01707.x

Cole, J.J., Prairie, Y.T., Caraco, N.F. et al. 2007. Plumbing the global carbon cycle: Integrating inland waters into the terrestrial carbon budget. *Ecosystems* 10, 1, 172–185. doi: 10.1007/s10021-006-9013-8

Conant, R.T., Ryan, M.G., Ågren, G.I. et al. 2011. Temperature and soil organic matter decomposition rates – Synthesis of current knowledge and a way forward. *Global Change Biology* 17, 11, 3392–3404. doi: 10.1111/j.1365-2486.2011.02496.x

Constantine, J.A., Dunne, T., Ahmed, J. et al. 2014. Sediment supply as a driver of river meandering and floodplain evolution in the Amazon Basin. *Nature Geoscience* 7, 899–903. doi: 10.1038/ngeo2282

Cook, J., Edwards, A., Takeuchi, N. et al. 2016. Cryoconite: The dark biological secret of the cryosphere. *Progress in Physical Geography* 40, 1, 66–111. doi: 10.1177/0309133315616574

Cooke, R.U. 1976. Urban geomorphology. *The Geographical Journal* 142, 1, 59–65. doi: 10.2307/1796025

Coombes, M.A. 2016. Biogeomorphology: Diverse, integrative and useful. *Earth Surface Processes and Landforms* 41, 15, 2296–2300. doi: 10.1002/esp.4055

Cooper, A.H., Brown, T.J., Price, S.J. et al. 2018. Humans are the most significant global geomorphological driving force of the 21st century. *The Anthropocene Review* 5, 3, 222–229. doi: 10.1177/2053019618800234

Corenblit, D., Baas, A., Balke, T. et al. 2015. Engineer pioneer plants respond to and affect geomorphic constraints similarly along water–terrestrial interfaces world-wide. *Global Ecology and Biogeography* 24, 12, 1363–1376. doi: 10.1111/geb.12373

Costanza, R., d'Arge, R., and de Groot, R. 1997. The value of the world's ecosystem services and natural capital. *Nature* 387, 253–260. doi: 10.1038/387253a0

Coulthard, T.J. and Van De Wiel, M.J. 2006. A cellular model of river meandering. *Earth Surface Processes and Landforms* 31, 1, 123–132. doi: 10.1002/esp.1315

Cowley, K.L., Fryirs, K.A., and Hose, G.C. 2016. Identifying key sedimentary indicators of geomorphic structure and function of upland swamps in the Blue Mountains for use in condition assessment and monitoring. *Catena* 147, 564–577. doi: 10.1016/j.catena.2016.08.016

Craft, C.B. and Casey, W.P. 2000. Sediment and nutrient accumulation in floodplain and depressional freshwater wetlands of Georgia, USA. *Wetlands* 20, 2, 323–332. doi: 10.1672/0277-5212(2000)020[0323:SANAIF]2.0.CO;2

Craft, J.A., Stanford, J.A., and Pusch, M. 2002. Microbial respiration within a floodplain aquifer of a large gravel-bed river. *Freshwater Biology* 47, 2, 251–261. doi: 10.1046/j.1365-2427.2002.00803.x

Cramer, W., Kicklighter, D.W., Bondeau, A. et al. 1999. Comparing global models of terrestrial net primary productivity (NPP): Overview and key results. *Global Change Biology* 5, S1, 1–15. doi: 10.1046/j.1365-2486.1999.00009.x

Croissant, T., Steer, P., Lague, D. et al. 2019. Seismic cycles, earthquakes, landslides and sediment fluxes: Linking tectonics to surface processes using a reduced-complexity model. *Geomorphology* 339, 87–103. doi: 10.1016/j.geomorph.2019.04.017

Crowe, S.K., Evans, M.G., and Allott, T.E.H. 2008. Geomorphological controls on the re-vegetation of erosion gullies in blanket peat: Implications for bog restoration. *Mires and Peat* 1, 1–14. http://mires-and-peat.net/pages/volumes/map03/map0301.php

Crowell, S.M.R., Kawa, S.R., Browell, E.V. et al. 2018. On the ability of space-based passive and active remote sensing observations of CO_2 to detect flux perturbations to the carbon cycle. *Journal of Geophysical Research Atmospheres* 123, 2, 1460–1477. doi: 10.1002/2017JD027836

Crump, B.C., Amaral-Zettler, L.A., and Kling, G.W. 2012. Microbial diversity in arctic freshwaters is structured by inoculation of microbes from soils. *Journal of the International Society for Microbial Biology* 6, 1629–1639. doi: 10.1038/ismej.2012.9

Crump, B.C., Hopkinson, C.S., Sogin, M.L. et al. 2004. Microbial biogeography along an estuarine salinity gradient: Combined influences of bacterial growth and residence time. *Applied and Environmental Microbiology* 70, 3, 1494–1505. doi: 10.1128/AEM.70.3.1494-1505.2004

Crutzen, P. and Stoermer, E.F. 2000. The "Anthropocene". IGBP Newsletter 41, 17–18.

Crutzen, P.J. 2002. Geology of mankind. *Nature* 415, 23. doi: 10.1038/415023a

Csima, P. 2010. Urban development and anthropogenic geomorphology, in Szabó, J., Dávid, L., and Lóczy, D. (eds.), *Anthropogenic Geomorphology*. Springer, Dordrecht, Netherlands, pp. 179–187. doi: 10.1007/978-90-481-3058-0_12

Cubasch, U., Wuebbles, D., and D. Chen et al. 2013. Introduction, in Stocker, T.F., Qin, D., Plattner, G.-K. et al.(eds.), *Climate Change 2013: The physical science basis*. Contribution of Working Group I to the 5th Assessment Report of the Intergovernmental Panel on Climate Change, 119–158. Cambridge University Press, Cambridge, UK and New York.

Cubizolle, H. and Thebaud-Sorger, M. 2014. A geographical model for an altitudinal zonation of mire types in the uplands of Western Europe: The example case being the Monts du Forez mountain range in France. *Mires and Peat* 1515, 1–16. http://www.mires-and-peat.net/pages/volumes/map15/map1502.php

Dai, M., Yin, Z., Meng, F. et al. 2012. Spatial distribution of riverine DOC inputs to the ocean: An updated global synthesis. *Current Opinion in Environmental Sustainability* 4, 2, 170–178. doi: 10.1016/j.cosust.2012.03.003

Darjany, L.E., Whitcraft, C.R., and Dillon, J.G. 2014. Lignocellulose-responsive bacteria in a southern California salt marsh identified by stable isotope probing. *Frontiers in Microbiology* 5, 1–9. doi: 10.3389/fmicb.2014.00263

Davidson, E.A. and Janssens, I.A. 2006. Temperature sensitivity of soil carbon decomposition and feedbacks to climate change. *Nature* 440, 165–173. doi: 10.1038/nature04514

Davies, P.M., Bunn, S.E., and Hamilton, S.K. 2008. Primary production in tropical streams and rivers, in Dudgeon, D. (ed.), *Tropical Stream Ecology*. Elsevier, London, UK, pp. 23–42.

Davies, T.R.H. and Korup, O. 2010. Sediment cascades in active landscapes, in Burt, T.P. and Allison, R.J. (eds.), *Sediment Cascades: An Integrated Approach*. Wiley Blackwell, Oxford, UK, pp. 89–115.

DeGrood, S.H., Claassen, V.P., and Scow, K.M. 2005. Microbial community composition on native and drastically disturbed serpentine soils. *Soil Biology and Biochemistry* 37, 8, 1427–1435. doi: 10.1016/j.soilbio.2004.12.013

De Reu, J., Bourgeois, J., Bats, M. et al. 2013. Application of the topographic position index to heterogeneous landscapes. *Geomorphology* 186, 39–49. doi: 10.1016/j.geomorph.2012.12.015

Deslogest, J.R. and Church, M. 1987. Channel and floodplain facies in a wandering gravel-bed river, in Ethridge, F.G., Flores, R.M., and Harvey, M.D. (eds.), *Recent Developments in Fluvial Sedimentology. Special Publications of the Society of Economic Paleontologists and Mineralogists* 39, 99–109.

de Vente, J., Poesen, J., Arabkhedri, M. et al. 2007. The sediment delivery problem revisited. *Progress in Physical Geography* 31, 2, 155–178. doi: 10.1177/0309133307076485

de Vries, F., Liiri, M., Bjørnlund, L. et al. 2012. Land use alters the resistance and resilience of soil food webs to drought. *Nature Climate Change* 2, 276–280. doi: 10.1038/nclimate1368

de Vries, F.T., Griffiths, R.I., Bailey, M. et al. 2018. Soil bacterial networks are less stable under drought than fungal networks. *Nature Communications* 9, 1–12. doi: 10.1038/s41467-018-05516-7

DeVries, P., Fetherston, K.L., Vitale, A. et al. 2012. Emulating riverine landscape controls of beaver in stream restoration. *Fisheries* 37, 6, 246–255. doi: 10.1080/03632415.2012.687263

Deyong, Y., Hongbo, S., Peijun, S. et al. 2009. How does the conversion of land cover to urban use affect net primary productivity? A case study in Shenzhen city, China. *Agricultural and Forest Meteorology* 149, 11, 2054–2060. doi: 10.1016/j.agrformet.2009.07.012

Dietrich, W.E. and Dunne, T. 1978. Sediment budget for a small catchment in mountainous terrain. *Zeitschrift für Geomorphologie Supplement* 29, 191–206. http://wpg.forestry.oregonstate.edu/sites/wpg/files/seminars/1978_Dietrich%26Dunne.pdf

Dietrich, W.E., Dunne, T., Humphrey, N.F. et al. (1982). Construction of sediment budgets for drainage basins, in Swanson, F.J., Janda, R.J., Dunne, T. et al. (eds.), *Sediment Budgets and Routing in Forested Drainage Basins.* General Technical Report PNW-141, US Forest Service, 5–23.

Dini-Andreote, F., Pylro, V.S., Baldrian, P. et al. 2016. Ecological succession reveals potential signatures of marine–terrestrial transition in salt marsh fungal communities. *Journal of the International Society for Microbial Ecology* 10, 1984–1997. doi: 10.1038/ismej.2015.254

Dixon, S.J., Viles, H.A., and Garrett, B.L. 2018. Ozymandias in the Anthropocene: The city as an emerging landform. *Area* 50, 1, 117–125. doi: 10.1111/area.12358

Doetterl, S., Berhe, A.A., Nadeu, E. et al. 2016. Erosion, deposition and soil carbon: A review of process-level controls, experimental tools and models to address C cycling in dynamic landscapes. *Earth-Science Reviews* 154, 102–122. doi: 10.1016/j.earscirev.2015.12.005

Dosseto, A., Bourdon, B. Gaillardet, J. et al. 2006a Weathering and transport of sediments in the Bolivian Andes: Time constraints from uranium-series isotopes. *Earth and Planetary Science Letters,* 248 3–4, 759–771. https://doi.org/10.1016/j.epsl.2006.06.027

Dosseto, A., Bouron, B., Gaillardet, J. et al.2006b. Time scale and conditions of weathering under tropical climate: Study of the Amazon basin with U-series. *Geochimica et Cosmochimica Acta* 70, 1, 71–89. https://doi.org/10.1016/j.gca.2005.06.033

Doughty, C.L., Langley, J.A., Walker, W.S. et al. 2016. Mangrove range expansion rapidly increases coastal wetland carbon storage. *Estuaries and Coasts* 39, 2, 385–396. doi: 10.1007/s12237-015-9993-8

Douglas, I., and Lawson, N. 2000. The human dimensions of geomorphological work in Britain. *Journal of Industrial Ecology* 4, 2, 9–33. doi: 10.1162/108819800569771

Duarte, C.M. 2002. The future of seagrass meadows. *Environmental Conservation* 29, 2, 192–206. doi: 10.1017/S0376892902000127

Dufour, S. and Piégay, H. 2009. From the myth of a lost paradise to targeted river restoration: Forget natural references and focus on human benefits. *River Research and Applications* 25, 5, 568–581. doi: 10.1002/rra.1239

Dungait, J.A.J., Ghee, C., Rowan, J.S. et al. 2013. Microbial responses to the erosional redistribution of soil organic carbon in arable fields. *Soil Biology and Biochemistry* 60, 195–201. doi: 10.1016/j.soilbio.2013.01.027

East, A.E., Pess, G.R., Bountry, J.A. et al. 2015. Large-scale dam removal on the Elwha River, Washington, USA: River channel and floodplain geomorphic change. *Geomorphology* 228, 765–786. doi: 10.1016/j.geomorph.2014.08.028

Edmondson, J.L., Davies, Z.G., McHugh, N. et al. 2012. Organic carbon hidden in urban ecosystems. *Nature Scientific Reports* 2, 1–7. doi: 10.1038/srep00963

Edmondson, J.L., Stott, I., Potter, J. et al. 2015. Black carbon contribution to organic carbon stocks in urban soil. *Environmental Science & Technology* 49, 14, 8339–8346. doi: 10.1021/acs.est.5b00313

Edwards, A., Pachebat, J.A., Swain, M. et al. 2013. A metagenomic snapshot of taxonomic and functional diversity in an alpine glacier cryoconite ecosystem. *Environmental Research Letters* 8, 3, 1–11. doi: 10.1088/1748-9326/8/3/035003

Edwards, C., Hales, B.A., Hall, G.H. et al. 1998. Microbiological processes in the terrestrial carbon cycle: Methane cycling in peat. *Atmospheric Environment* 32, 19, 3247–3255. doi: 10.1016/S1352-2310(98)00107-1

Ekschmitt, K., Kandeler, E., Poll, C. et al. 2008. Soil-carbon preservation through habitat constraints and biological limitations on decomposer activity. *Journal of Plant Nutrition and Soil Science* 171, 1, 27–35. doi: 10.1002/jpln.200700051

Elliott, D.R., Caporn, S.J.M., Nwaishi, F. et al. 2015. Bacterial and fungal communities in a degraded ombrotrophic peatland undergoing natural and managed re-vegetation. *PLOS ONE* 10, 5, 1–20. doi: 10.1371/journal.pone.0124726

Evans, C., Allott, T., Billett, M. et al. 2013. Greenhouse gas emissions associated with non gaseous losses of carbon from peatlands: Fate of particulate and dissolved carbon. DEFRA project SP1205 Final report.

Evans, C.D., Page, S.E., Jones, T. et al. 2014a. Contrasting vulnerability of drained tropical and high-latitude peatlands to fluvial loss of stored carbon. *Global Biogeochemical Cycles* 28, 11, 1215–1234. doi: 10.1002/2013gb004782

Evans, C.D., Bonn, A., Holden, J. et al. 2014b. Relationships between anthropogenic pressures and ecosystem functions in UK blanket bogs: Linking process understanding to ecosystem service valuation. *Ecosystem Services* 9, 5–19. doi: 10.1016/j.ecoser.2014.06.013

Evans, D. 2014. *Glacial Landsystems*. Routledge, London.

Evans, D.J.A., Lemmen, D.S., and Rea, B.R. 1999. Glacial landsystems of the southwest Laurentide Ice Sheet: Modern Icelandic analogues. *Journal of Quaternary Science* 14, 7, 673–691. doi: 10.1002/(SICI)1099-1417(199912)14:7<673::AID-JQS467>3.0.CO;2-#

Evans, M. 1997. Temporal and spatial representativeness of alpine sediment yields: Cascade Mountains, British Columbia. *Earth Surface Processes and Landforms* 22, 3, 287–295. doi: 10.1002/(sici)1096-9837(199703)22:3%3C287::aid-esp757%3E3.0.co;2-v

Evans, M. and Lindsay, J. 2010a. High resolution quantification of gully erosion in upland peatlands at the landscape scale. *Earth Surface Processes and Landforms* 35, 8, 876–886. doi: 10.1002/esp.1918

Evans, M. and Lindsay, J. 2010b. Impact of gully erosion on carbon sequestration in blanket peatlands. *Climate Research* 45, 1, 31–41. doi: 10.3354/cr00887

Evans, M., Quine, T., and Kuhn, N. 2013. Geomorphology and terrestrial carbon cycling. *Earth Surface Processes and Landforms* 38, 1, 103–105. doi: 10.1002/esp.3337

Evans, M. and Warburton, J. 2007. *Geomorphology of Upland Peat: Erosion, Form and Landscape Change*. John Wiley & Sons, Chichester, UK.

Evans, M., Warburton, J., and Yang, J. 2006. Eroding blanket peat catchments: Global and local implications of upland organic sediment budgets. *Geomorphology* 79, 1–2, 45–57. doi: 10.1016/j.geomorph.2005.09.015

Fatichi, S., Pappas, C., Zscheischler, J. et al. 2019. Modelling carbon sources and sinks in terrestrial vegetation. *New Phytologist* 221, 2, 652–668. doi: 10.1111/nph.15451

Fellman, J.B., Hood, E., Raymond, P.A. et al. 2015. Evidence for the assimilation of ancient glacier organic carbon in a proglacial stream food web. *Limnology and Oceanography* 60, 4, 1118–1128. doi: 10.1002/lno.10088

Ferguson, R. 1981. Channel forms and channel changes, in Lewin, J. (ed.), *British Rivers*. Allen & Unwin, London, UK, pp. 90–125.

Fernandes, J.P., Almeida, C.M.R., Andreotti, F. et al. 2017. Response of microbial communities colonizing salt marsh plants rhizosphere to copper oxide nanoparticles contamination and its implications for phytoremediation processes. *Science of the Total Environment* 581–582, 801–810. doi: 10.1016/j.scitotenv.2017.01.015

Foght, J., Aislabie, J., Turner, S. et al. 2004. Culturable bacteria in subglacial sediments and ice from two southern hemisphere glaciers. *Microbial Ecology* 47, 4, 329–340. doi: 10.1007/s00248-003-1036-5

Foley, J.A., DeFries, R., Asner, G.P. et al. 2005. Global consequences of land use. *Science* 309, 5734, 570–574. doi: 10.1126/science.1111772

Foster, G.L. and Vance, D. 2006. Negligible glacial–interglacial variation in continental chemical weathering rates. *Nature* 444, 918–921. doi: 10.1038/nature05365

Fourqurean, J.W., Duarte, C.M., Kennedy, H. et al. 2012. Seagrass ecosystems as a globally significant carbon stock. *Nature Geoscience* 5, 505–509. doi: 10.1038/ngeo1477

Freeman, C., Ostle, N., Fenner, N. et al. 2004. A regulatory role for phenol oxidase during decomposition in peatlands. *Soil Biology and Biochemistry* 36, 10, 1663–1667. doi: 10.1016/j.soilbio.2004.07.012

Freidman, B.L. and Fryirs, K.A. 2015. Rehabilitating upland swamps using environmental histories: A case study of the Blue Mountains peat swamps, eastern Australia. *Geografiska Annaler: Series A Physical Geography* 97, 2, 337–353. doi: 10.1111/geoa.12068

Friedrichs, C.T. and Perry, J.E. 2001. Tidal salt marsh morphodynamics: A synthesis. *Journal of Coastal Research* 27, 7–37. https://www.jstor.org/stable/25736162

Frings, P.J., Clymans, W., Fontorbe, G. et al. 2015. Silicate weathering in the Ganges alluvial plain. *Earth and Planetary Science Letters* 427, 136–148. doi: 10.1016/j.epsl.2015.06.049

Frith, N.V., Hilton, R.G., Howarth, J.D. et al. 2018. Carbon export from mountain forests enhanced by earthquake-triggered landslides over millennia. *Nature Geoscience* 11, 772–776. doi: 10.1038/s41561-018-0216-3

Fritz, M., Vonk, J.E., and Lantuit, H. 2017. Collapsing Arctic coastlines. *Nature Climate Change* 7, 6–7. doi: 10.1038/nclimate3188

Fryirs, K. 2013. (Dis)Connectivity in catchment sediment cascades: A fresh look at the sediment delivery problem. *Earth Surface Processes and Landforms* 38, 1, 30–46. doi: 10.1002/esp.3242

Fryirs, K.A. 2015. Developing and using geomorphic condition assessments for river rehabilitation planning, implementation and monitoring. *WIRES Water* 2, 6, 649–667. doi: 10.1002/wat2.1100

Fryirs, K.A., Cowley, K., and Hose, G.C. 2016. Intrinsic and extrinsic controls on the geomorphic condition of upland swamps in Eastern NSW. *Catena* 137, 100–112. doi: 10.1016/j.catena.2015.09.002

Gacia, E., Duarte, C.M., Marbà, N. et al. 2003. Sediment deposition and production in SE-Asia seagrass meadows. *Estuarine, Coastal and Shelf Science* 56, 5–6, 909–919. doi: 10.1016/S0272-7714(02)00286-X

Gaillardet, J., Dupré, B., Louvat, P. et al. 1999. Global silicate weathering and CO_2 consumption rates deduced from the chemistry of large rivers. *Chemical Geology* 159, 1–4, 3–30. doi: 10.1016/S0009-2541(99)00031-5

Galy, V. and Eglinton, T. 2011. Protracted storage of biospheric carbon in the Ganges–Brahmaputra basin. *Nature Geoscience* 4, 843–847. doi: 10.1038/ngeo1293

Galy, V., Peucker-Ehrenbrink, B., and Eglinton, T. 2015. Global carbon export from the terrestrial biosphere controlled by erosion. *Nature* 521, 204–207. doi: 10.1038/nature14400

Ganju, N.K., Defne, Z., Kirwan, M.L. et al. 2017. Spatially integrative metrics reveal hidden vulnerability of microtidal salt marshes. *Nature Communications* 8, 1–7. doi: 10.1038/ncomms14156

Gibson, H.S., Worrall, F., Burt, T.P. et al. 2009. DOC budgets of drained peat catchments: Implications for DOC production in peat soils. *Hydrological Processes* 23, 13, 1901–1911. doi: 10.1002/hyp.7296

Gilbert, D., Amblard, C., Bourdier, G. et al. 1998. The microbial loop at the surface of a peatland: Structure, function, and impact of nutrient input. *Microbial Ecology* 35, 1, 83–93. https://www.jstor.org/stable/4251541

Gilbert, D., and Mitchell, E.A.D. 2006. Microbial diversity in Sphagnum peatlands, in Martini, I.P., Martinez Cortizas, A., and Chesworth, W. (eds.), *Peatlands: Evolution and Records of Environmental and Climate Changes.* Elsevier, Amsterdam, pp. 287–318.

Gilbert, G.K. 1877. *Geology of the Henry Mountains.* U.S. Geographical and Geological Survey, Washington, DC, 160pp.

Girardin, C.A.J., Malhi, Y., Aragão, L.E.O.C. et al. 2010. Net primary productivity allocation and cycling of carbon along a tropical forest elevational transect in the Peruvian Andes: Net primary productivity from Andes to Amazon. *Global Change Biology* 16, 12, 3176–3192. doi: 10.1111/j.1365-2486.2010.02235.x

Giri, C., Ochieng, E., Tieszen, L.L. et al. 2011. Status and distribution of mangrove forests of the world using earth observation satellite data. *Global Ecology and Biogeography* 20, 1, 154–159. doi: 10.1111/j.1466-8238.2010.00584.x

Glendell, M., McShane, G., Farrow, L. et al. 2017. Testing the utility of structure-from-motion photogrammetry reconstructions using small unmanned aerial vehicles and ground photography to estimate the extent of upland soil erosion. *Earth Surface Processes and Landforms* 42, 12, 1860–1871. doi: 10.1002/esp.4142

Glymph, L.M. 1954. Studies of sediment yields from watersheds. *International Association of Hydrology Publications* 36, 173–191.

Gold, A.J., Addy, K., Morrison, A. et al. 2016. Will dam removal increase nitrogen Flux to Estuaries? *Water* 8, 11, 1–16. doi: 10.3390/w8110522

Golovchenko, A., Tikhonova, E., and Zvyagintsev, D. 2007. Abundance, biomass, structure, and activity of the microbial complexes of minerotrophic and ombrotrophic peatlands. *Microbiology* 76, 5, 630–637. doi: 10.1134/S0026261707050177

Gomez, B., Baisden, W.T., and Rogers, K.M. 2010. Variable composition of particle-bound organic carbon in steepland river systems. *Journal of Geophysical Research* 115, F4, 1–9. doi: 10.1029/2010JF001713

Gomez, B., Eden, D.N., Peacock, D.H. et al. 1998. Floodplain construction by recent, rapid vertical accretion: Waipaoa River, New Zealand. *Earth Surface Processes and Landforms* 23, 5, 405–413. doi: 10.1002/(SICI)1096-9837(199805)23:5<405::AID-ESP854>3.0.CO;2-X

Gomez, B. and Trustrum, N.A. 2005. Landscape disturbance and organic carbon in alluvium bordering steepland rivers, east coast continental margin, New Zealand, in Garcia, C. and Batalla, R.J. (eds.), *Developments in Earth Surface Processes, Catchment Dynamics and River Processes Mediterranean and Other Climate Regions.* Elsevier, Amsterdam, The Netherlands, pp. 103–116.

Gomez, B., Trustrum, N.A., Hicks, D.M. et al. 2003. Production, storage, and output of particulate organic carbon: Waipaoa River basin, New Zealand: Production of particulate organic carbon. *Water Resources Research* 39, 6, 1–8. doi: 10.1029/2002WR001619

Gomi, T., Sidle, R.C., and Richardson, J.S. 2002. Understanding processes and downstream linkages of headwater systems. *BioScience* 52, 10, 905–916. doi: 10.1641/0006-3568(2002)052[0905:UPADLO]2.0.CO;2

Gorham, E. 1991. Northern peatlands: Role in the carbon cycle and probable responses to climatic warming. *Ecological Applications* 1, 2, 182–195. doi: 10.2307/1941811

Goudie, A.S. 2013. *The Human Impact on the Natural Environment: Past, Present, and Future.* John Wiley & Sons, Cambridge, UK.

Goudie, A.S., and Viles, H.A. 2012. Weathering and the global carbon cycle: Geomorphological perspectives. *Earth-Science Reviews* 113, 1–2, 59–71. doi: 10.1016/j.earscirev.2012.03.005

Goudie, A.S. and Viles, H.A. 2016. *Geomorphology in the Anthropocene.* Cambridge University Press, Cambridge, UK.

Grabowski, R.C., Gurnell, A.M., Burgess-Gamble, L. et al. 2019. The current state of the use of large wood in river restoration and management. *Water and Environment Journal* 33, 3, 366–377. doi: 10.1111/wej.12465

Grace, J. 2001. Carbon cycle, in Levin, S.A. (ed.), *Encyclopedia of Biodiversity.* Elsevier, London, pp. 609–628.

Graves, C.J., Makrides, E.J., Schmidt, V.T. et al. 2016. Functional responses of salt marsh microbial communities to long-term nutrient enrichment. *Applied and Environmental Microbiology* 82, 2862–2871. doi: 10.1128/AEM.03990-15

Greiner, J.T., McGlathery, K.J., Gunnell, J. et al. 2013. Seagrass restoration enhances "blue carbon" sequestration in coastal waters. *PLOS ONE* 8, 8, 1–8. doi: 10.1371/journal.pone.0072469

Grubaugh, J.W. and Anderson, R.V. 1989. Upper Mississippi River: Seasonal and floodplain forest influences on organic matter transport. *Hydrobiologia* 174, 3, 235–244. doi: 10.1007/BF00008163

Guelland, K., Hagedorn, F., Smittenberg, R.H. et al. 2013. Evolution of carbon fluxes during initial soil formation along the forefield of Damma glacier, Switzerland. *Biogeochemistry* 113, 1–3, 545–561. doi: 10.1007/s10533-012-9785-1

Guo, L., Ping, C.L., and Macdonald, R.W. 2007. Mobilization pathways of organic carbon from permafrost to arctic rivers in a changing climate. *Geophysical Research Letters* 34, 13, 1–5. doi: 10.1029/2007GL030689

Gurnell, A., Lee, M., and Souch, C. 2007. Urban rivers: Hydrology, geomorphology, ecology and opportunities for change. *Geography Compass* 1, 5, 1118–1137. doi: 10.1111/j.1749-8198.2007.00058.x

Gurnell, A.M. 1998. The hydrogeomorphological effects of beaver dam-building activity. *Progress in Physical Geography: Earth and Environment* 22, 2, 167–189. doi: 10.1177/030913339802200202

Gurwick, N.P., Groffman, P.M., Yavitt, J.B. et al. 2008. Microbially available carbon in buried riparian soils in a glaciated landscape. *Soil Biology and Biochemistry* 40, 1, 85–96. doi: 10.1016/j.soilbio.2007.07.007

Gutknecht, J.L.M., Field, C.B., and Balser, T.C. 2012. Microbial communities and their responses to simulated global change fluctuate greatly over multiple years. *Global Change Biology* 18, 7, 2256–2269. doi: 10.1111/j.1365-2486.2012.02686.x

Guyette, R.P., Cole, W.G., Dey, D.C. et al. 2002. Perspectives on the age and distribution of large wood in riparian carbon pools. *Canadian Journal of Fisheries and Aquatic Sciences* 59, 3, 578–585. doi: 10.1139/f02-026

Guyette, R.P., Dey, D.C., and Stambaugh, M.C. 2008. The temporal distribution and carbon storage of large oak wood in streams and floodplain deposits. *Ecosystems* 11, 643–653. doi: 10.1007/s10021-008-9149-5

Hack, J.T. 1957. *Studies of Longitudinal Profiles in Virginia and Maryland*. USGS Professional Papers 294-B, 45–97, United States Department Government Printing Office, Washington, DC.

Hall, S. and Hopkins, D.W. 2015. A microbial biomass and respiration of soil, peat and decomposing plant litter in a raised mire. *Plant, Soil and Environment* 61, 9, 405–409. doi: 10.17221/311/2015-PSE

Ham, D. and Church, M. 2000. Bed-material transport estimated from channel morphodynamics: Chliwack river, British Columbia. *Earth Surface Processes and Landforms* 25, 1123–1142. https://doi.org/10.1002/1096-9837(200009)25:10<1123::AID-ESP122>3.0.CO;2-9

Harden, J.W., Sharpe, J.M., Parton, W.J. et al. 1999. Dynamic replacement and loss of soil carbon on eroding cropland. *Global Biogeochemical Cycles* 13, 4, 885–901. doi: 10.1029/1999GB900061

Harvey, A.M. 2001. Coupling between hillslopes and channels in upland fluvial systems: Implications for landscape sensitivity, illustrated from the Howgill Fells, northwest England. *Catena* 42, 2–4, 225–250. doi: 10.1016/S0341-8162(00)00139-9

Hawkes, C.V., Waring, B.G., Rocca, J.D. et al. 2017. Historical climate controls soil respiration responses to current soil moisture. *Proceedings of the National Academy of Sciences of the United States of America* 114, 24, 6322–6327. doi: 10.1073/pnas.1620811114

Heckman, T. and Schwanghart, W. 2013. Geomorphic coupling and sediment connectivity in an alpine catchment – Exploring sediment cascades using graph theory. *Geomorphology* 182, 89–103. https://doi.org/10.1016/j.geomorph.2012.10.033

Hedges, J.I., Keil, R.G., and Benner, R. 1997. What happens to terrestrial organic matter in the ocean? *Organic Geochemistry* 27, 5–6, 195–212. doi: 10.1016/S0146-6380(97)00066-1

Hein, T., Baranyi, C., Herndl, G.J. et al. 2003. Allochthonous and autochthonous particulate organic matter in floodplains of the River Danube: The importance of hydrological connectivity. *Freshwater Biology* 48, 2, 220–232. doi: 10.1046/j.1365-2427.2003.00981.x

Hemingway, J.D., Hilton, R.G., Hovius, N. et al.2018. Microbial oxidation of lithospheric organic carbon in rapidly eroding tropical mountain soils. *Science* 36, 6385, 209–212. doi: 10.1126/science.aao6463

Hemingway, J.D., Rothman, D.H., Grant, K.E. et al.2019. Mineral protection regulates long-term global preservation of natural organic carbon. *Nature* 570, 228–231. https://doi.org/10.1038/s41586-019-1280-6

Hernes, P.J. and Benner, R. 2003. Photochemical and microbial degradation of dissolved lignin phenols: Implications for the fate of terrigenous dissolved organic matter in marine environments. *Journal of Geophysical Research: Oceans* 108, C9, 1–9. doi: 10.1029/2002JC001421

Herrmann, M., Najjar, R.G., Kemp, W.M. et al. 2015. Net ecosystem production and organic carbon balance of U.S. East Coast estuaries: A synthesis approach. *Global Biogeochemical Cycles* 29, 1, 96–111. doi: 10.1002/2013GB004736

Hevia, J.N., de Araújo, J.C., and Manso, J.M. 2014. Assessment of 80 years of ancient-badlands restoration in Saldaña, Spain. *Earth Surface Processes and Landforms* 39, 12, 1563–1575. doi: 10.1002/esp.3541

Hewitt, K. 2006. Disturbance regime landscapes: mountain drainage systems interrupted by large rockslides. *Progress in Physical Geography: Earth and Environment* 30, 3, 365–393. doi:10.1191/0309133306pp486ra

Hickin, E.J. and Nanson, G.C. 1984. Lateral migration rates of river bends. *Journal of Hydraulic Engineering* 110, 11, 1557–1567. doi: 10.1061/(ASCE)0733-9429(1984)110:11(1557)

Hilton, R.G. 2017. Climate regulates the erosional carbon export from the terrestrial biosphere. *Geomorphology* 277, 118–132. doi: 10.1016/j.geomorph.2016.03.028

Hilton, R.G., Galy, A., and Hovius, N. 2008. Riverine particulate organic carbon from an active mountain belt: Importance of landslides. *Global Biogeochemical Cycles* 22, 1, 1–12. doi: 10.1029/2006GB002905

Hilton, R.G., Galy, A., Hovius, N. et al. 2011a. Efficient transport of fossil organic carbon to the ocean by steep mountain rivers: An orogenic carbon sequestration mechanism. *Geology* 39, 1, 71–74. doi: 10.1130/G31352.1

Hilton, R.G., Meunier, P., Hovius, N. et al. 2011b. Landslide impact on organic carbon cycling in a temperate montane forest. *Earth Surface Processes and Landforms* 36, 12, 1670–1679. doi: 10.1002/esp.2191

Hilton, R.G., Galy, V., Gaillardet, J. et al. 2015. Erosion of organic carbon in the Arctic as a geological carbon dioxide sink. *Nature* 524, 84–87. doi: 10.1038/nature14653

Hilton, R.G. and West, A.J. 2020. Mountains, erosion and the carbon cycle. *Nature Reviews Earth & Environment* 1, 284–299. doi: 10.1038/s43017-020-0058-6

Hodson, A., Anesio, A.M., Ng, F. et al. 2007. A glacier respires: Quantifying the distribution and respiration CO_2 flux of cryoconite across an entire Arctic supraglacial ecosystem: Respiration rates upon an Arctic glacier. *Journal of Geophysical Research: Biogeosciences* 112, G4, 1–9. doi: 10.1029/2007JG000452

Hoffmann, T. 2015. Sediment residence time and connectivity in non-equilibrium and transient geomorphic systems. *Earth-Science Reviews* 150, 609–627. doi: 10.1016/j.earscirev.2015.07.008

Hoffmann, T., Glatzel, S., and Dikau, R. 2009. A carbon storage perspective on alluvial sediment storage in the Rhine catchment. *Geomorphology* 108, 1–2, 127–137. doi: 10.1016/j.geomorph.2007.11.015

Hoffmann, T., Mudd, S.M., van Oost, K. et al. 2013a. Short Communication: Humans and the missing C-sink: Erosion and burial of soil carbon through time. *Earth Surface Dynamics* 1, 45–52. doi: 10.5194/esurf-1-45-2013

Hoffmann, T., Schlummer, M., Notebaert, B. et al. 2013b. Carbon burial in soil sediments from Holocene agricultural erosion, Central Europe: Carbon burial in sediments. *Global Biogeochemical Cycles* 27, 3, 828–835. doi: 10.1002/gbc.20071

Hoffmann, T., Thorndycraft, V.R., Brown, A.G. et al. 2010. Human impact on fluvial regimes and sediment flux during the Holocene: Review and future research agenda. *Global and Planetary Change* 72, 3, 87–98. doi: 10.1016/j.gloplacha.2010.04.008

Högberg, P., Nordgren, A., Buchmann, N. et al. 2001. Large-scale forest girdling shows that current photosynthesis drives soil respiration. *Nature* 411, 789–792. doi: 10.1038/35081058

Holden, J. 2005. Peatland hydrology and carbon release: Why small-scale process matters. *Philosophical Transactions of the Royal Society A: Mathematical Physical and Engineering Sciences* 363, 1837, 2891–2913. doi: 10.1098/rsta.2005.1671

Holden, J. and Burt, T.P. 2003. Hydrological studies on blanket peat: The significance of the acrotelm-catotelm model. *Journal of Ecology* 91, 1, 86–102. doi: 10.1046/j.1365-2745.2003.00748.x

Holden, J., Chapman, P.J., and Labadz, J.C. 2004. Artificial drainage of peatlands: Hydrological and hydrochemical process and wetland restoration. *Progress in Physical Geography* 28, 95–123. doi: 10.1191/0309133304pp403ra

Holl, K.D. and Aide, T.M. 2011. When and where to actively restore ecosystems? *Forest and Ecology Management* 261, 10, 1558–1563. doi: 10.1016/j.foreco.2010.07.004

Hønsvall, B.K. and Robertson, L.J. 2017. From research lab to standard environmental analysis tool: Will NASBA make the leap? *Water Research* 109, 389–397. doi: 10.1016/j.watres.2016.11.052

Hood, E., Battin, T.J., Fellman, J. et al. 2015. Storage and release of organic carbon from glaciers and ice sheets. *Nature Geoscience* 8, 91–96. doi: 10.1038/ngeo2331

Hood, E., Fellman, J., Spencer, R.G.M. et al. 2009. Glaciers as a source of ancient and labile organic matter to the marine environment. *Nature* 462, 1044–1047. doi: 10.1038/nature08580

Hooke, J.M. 1980. Magnitude and distribution of rates of river bank erosion. *Earth Surface Processes and Landforms* 5, 2, 143–157. doi: 10.1002/esp.3760050205

Horan, K., Hilton, R.G., Selby, D. et al. 2017. Mountain glaciation drives rapid oxidation of rock-bound organic carbon. *Science Advances* 3, 10, e1701107. doi: 10.1126/sciadv.1701107

Houben, P., Schmidt, M., Mauz, B. et al. 2013. Asynchronous Holocene colluvial and alluvial aggradation: A matter of hydrosedimentary connectivity. *The Holocene* 23, 4, 544–555. doi: 10.1177/0959683612463105

Hovius, N., Galy, A., Hilton, R.G. et al. 2011. Erosion-driven drawdown of atmospheric carbon dioxide: The organic pathway. *Applied Geochemistry* 26, S285–S287. doi: 10.1016/j.apgeochem.2011.03.082

Howes, B.L., Dacey, J.W.H., and King, G.M. 1984. Carbon flow through oxygen and sulfate reduction pathways in salt marsh sediments. *Limnology and Oceanography* 29, 5, 1037–1051. doi: 10.4319/lo.1984.29.5.1037

Hu, Y., Wang, L., Tang, Y. et al. 2014. Variability in soil microbial community and activity between coastal and riparian wetlands in the Yangtze River estuary – Potential impacts on carbon sequestration. *Soil Biology and Biochemistry* 70, 221–228. doi: 10.1016/j.soilbio.2013.12.025

Huang, T.-H., Fu, Y.-H., Pan, P.-Y. et al. 2012. Fluvial carbon fluxes in tropical rivers. *Current Opinion in Environmental Sustainability* 4, 2, 162–169. doi: 10.1016/j.cosust.2012.02.004

Hudson, P.F.,and Kesel, R.H. 2000. Channel migration and meander-bend curvature in the lower Mississippi River prior to major human modification. *Geology* 28, 6, 531–534. doi: 10.1130/0091-7613(2000)28<531:CMAMCI>2.0.CO;2

Huggett, R.J. 2003. *Fundamentals of Geomorphology.* Routledge, London, UK.

Hume, T.M., Snelder, T., Weatherhead, M. et al. 2007. A controlling factor approach to estuary classification. *Ocean and Coastal Management* 50, 11–12, 905–929. doi: 10.1016/j.ocecoaman.2007.05.009

Hyde, K.D. 1990. A comparison of the intertidal mycota of five mangrove tree species. *Asian Marine Biology* 7, 93–107.

Hyde, K.D. and Lee, S.Y. 1995. Ecology of mangrove fungi and their role in nutrient cycling: What gaps occur in our knowledge? *Hydrobiologia* 295, 1–3, 107–118. doi: 10.1007/BF00029117

Hyndes, G.A., Nagelkerken, I., McLeod, R.J. et al. 2014. Mechanisms and ecological role of carbon transfer within coastal seascapes. *Biological Reviews* 89, 1, 232–254. doi: 10.1111/brv.12055

Ingram, H.A.P. 1982. Size and shape in raised mire ecosystems: A geophysical model. *Nature* 297, 300–303. doi: 10.1038/297300a0

Ito, A. 2007. Simulated impacts of climate and land-cover change on soil erosion and implication for the carbon cycle, 1901 to 2100. *Geophysical Research Letters* 34, 9, 1–5. doi: 10.1029/2007GL029342

Ito, A. 2020. Constraining size-dependence of vegetation respiration rates. *Scientific Reports* 10, 1, 4304. doi: 10.1038/s41598-020-61239-0

Jacinthe, P.A. and Lal, R. 2001. A mass balance approach to assess carbon dioxide evolution during erosional events. *Land Degradation and Development* 12, 4, 329–339. doi: 10.1002/ldr.454

Jaffé, R., Boyer, J.N., Lu, X. et al. 2004. Source characterization of dissolved organic matter in a subtropical mangrove-dominated estuary by fluorescence analysis. *Marine Chemistry* 84, 3–4, 195–210. doi: 10.1016/j.marchem.2003.08.001

Jian, J., Steele, M.K., Thomas, R.Q. et al. 2018. Constraining estimates of global soil respiration by quantifying sources of variability. *Global Change Biology* 24, 9, 4143–4159. doi: 10.1111/gcb.14301

Jobbágy, E.G. and Jackson, R.B. 2000. The vertical distribution of soil organic carbon and its relation to climate and vegetation. *Ecological Applications* 10, 2, 423–436. doi: 10.1890/1051-0761(2000)010[0423:TVDOSO]2.0.CO;2

Jobson, H.E. 2001. Predicting river travel time from hydraulic characteristics. *Journal of Hydraulic Engineering* 127, 11, 911–918. doi: 10.1061/(ASCE)0733-9429(2001)127:11(911)

Johnston, C.A. 2014. Beaver pond effects on carbon storage in soils. *Geoderma* 213, 371–378. doi: 10.1016/j.geoderma.2013.08.025

Joosten, H. 2016. Peatlands across the globe, in Bonn, A., Allott, T.E.H., Evans, M. et al. (eds.), *Peatland Restoration and Ecosystem Services: Science Policy and Practice.* Cambridge University Press, Cambridge, UK, pp. 19–43.

Joosten, H., Sirin, A., Couwenberg, J. et al. 2016. The role of peatlands in climate regulation, in Bonn, A., Allott, T.E.H., and Evans, M. et al. (eds.), *Peatland Restoration and Ecosystem Services: Science Policy and Practice.* Cambridge University Press, Cambridge, UK, pp. 63–76.

Jörgensen, R.G., Raubuch, M., and Brandt, M. 2002. Soil microbial properties down the profile of a black earth buried by colluvium. *Journal of Plant Nutrition and Soil Science* 165, 3, 274–280. doi: 10.1002/1522-2624(200206)165:3<274::AID-JPLN274>3.0.CO;2-2

Joyce, H.M., Hardy, R.J., Warburton, J. et al. 2018. Sediment continuity through the upland sediment cascade: Geomorphic response of an upland river to an extreme flood event. *Geomorphology* 317, 45–61. doi: 10.1016/j.geomorph.2018.05.002

Junk, W.J., Bayley, P.B., and Sparks, R.E. 1989. The flood pulse concept in river-floodplain systems. Proceedings of the International Large River Symposium. *Canadian Special Publication of Fisheries and Aquatic Sciences* 106: 110–127.

Kabala, C.,and Zapart, J. 2012. Initial soil development and carbon accumulation on moraines of the rapidly retreating Werenskiold Glacier, SW Spitsbergen, Svalbard archipelago. *Geoderma* 175–176, 9–20. doi: 10.1016/j.geoderma.2012.01.025

Kaczmarek, Ł., Jakubowska, N., Celewicz-Gołdyn, S. et al. 2016. The microorganisms of cryoconite holes (algae, Archaea, bacteria, cyanobacteria, fungi, and Protista): A review. *Polar Record* 52, 2, 176–203. doi: 10.1017/S0032247415000637

Kaiser, K., Guggenberger, G., Haumaier, L. et al. 1997. Dissolved organic matter sorption on sub soils and minerals studied by ^{13}C-NMR and DRIFT spectroscopy. *European Journal of Soil Science* 48, 301–310. doi: 10.1111/j.1365-2389.1997.tb00550.x

Kamal, S. and Varma, A. 2008. Peatland microbiology, in Dion, P. and Nautiya, C.S. (eds.), *Microbiology of Extreme Soils.* Springer-Verlag, Berlin, pp. 177–203.

Kang, H. and Stanley, E.H. 2005. Effects of levees on soil microbial activity in a large river floodplain. *River Research and Applications* 21, 1, 19–25. doi: 10.1002/rra.811

Kang, S., Doh, S., Lee, D. et al. 2003. Topographic and climatic controls on soil respiration in six temperate mixed-hardwood forest slopes, Korea. *Global Change Biology* 9, 10, 1427–1437. doi: 10.1046/j.1365-2486.2003.00668.x

Kashian, D.M., Romme, W.H., Tinker, D.B. et al. 2006. Carbon storage on landscapes with stand-replacing fires. *BioScience* 56, 7, 598–606. doi: 10.1641/0006-3568(2006)56[598:CSOLWS]2.0.CO;2

Kaštovská, K., Elster, J., Stibal, M. et al. 2005. Microbial assemblages in soil microbial succession after glacial retreat in Svalbard (High Arctic). *Microbial Ecology* 50, 396–407. doi: 10.1007/s00248-005-0246-4

Kaštovská, K., Stibal, M., Šabacká, M. et al. 2007. Microbial community structure and ecology of subglacial sediments in two polythermal Svalbard glaciers characterized by epifluorescence microscopy and PLFA. *Polar Biology* 30, 3, 277–287. doi: 10.1007/s00300-006-0181-y

Kaye, J.P., McCulley, R.L., and Burke, I.C. 2005. Carbon fluxes, nitrogen cycling, and soil microbial communities in adjacent urban, native and agricultural ecosystems. *Global Change Biology* 11, 4, 575–587. doi: 10.1111/j.1365-2486.2005.00921.x

Keller, C.K. and Bacon, D.H. 1998. Soil respiration and georespiration distinguished by transport analyses of vadose CO_2, $^{13}CO_2$, and $^{14}CO_2$. *Global Biogeochemical Cycles* 12, 361–372. doi: 10.1029/98GB00742

Kelleway, J.J., Saintilan, N., Macreadie, P.I. et al. 2016a. Sedimentary factors are key predictors of carbon storage in SE Australian saltmarshes. *Ecosystems* 19, 5, 865–880. doi: 10.1007/s10021-016-9972-3

Kelleway, J.J., Saintilan, N., Macreadie, P.I. et al. 2016b. Seventy years of continuous encroachment substantially increases 'blue carbon' capacity as mangroves replace intertidal salt marshes. *Global Change Biology* 22, 3, 1097–1109. doi: 10.1111/gcb.13158

Kesel, R.H., Yodis, E.G., and McCraw, D.J. 1992. An approximation of the sediment budget of the lower mississippi river prior to major human modification. *Earth Surface Processes and Landforms* 17, 7, 711–722. doi: 10.1002/esp.3290170707

Kirchman, D.L. 2012. *Processes in Microbial Ecology*. Oxford University Press, Oxford, UK.

Kirkels, F.M.S.A., Cammeraat, L.H., and Kuhn, N.J. 2014. The fate of soil organic carbon upon erosion, transport and deposition in agricultural landscapes – A review of different concepts. *Geomorphology* 226, 94–105. doi: 10.1016/j.geomorph.2014.07.023

Kirschbaum, M.U.F. 1995. The temperature dependence of soil organic matter decomposition, and the effect of global warming on soil organic C storage. *Soil Biology and Biochemistry* 27, 6, 753–760. doi: 10.1016/0038-0717(94)00242-S

Kirschbaum, M.U.F. 2006. The temperature dependence of organic matter decomposition – still a topic of debate. *Soil Biology and Biochemistry* 38, 2510–2518. https://doi.org/10.1016/j.soilbio.2006.01.030

Kløve, B. 1998. Erosion and sediment delivery from peat mines. *Soil and Tillage Research* 45, 1–2, 199–216. doi: 10.1016/S0933-3630(97)00018-4

Knighton, D. 1998. *Fluvial Forms and Processes*. Arnold, London.

Knighton, A.D. 1999. Downstream variation in stream power. *Geomorphology* 29, 3–4, 293–306. doi: 10.1016/S0169-555X(99)00015-X

Knox, J. 2006. Floodplain sedimentation in the upper Mississippi valley: Natural versus human accelerated. *Geomorphology* 79, 3–4, 286–310. doi: 10.1016/j.geomorph.2006.06.031

Kobayashi, Y., Ryder, D., Gordon, G. et al. 2009. Short-term response of nutrients, carbon and planktonic microbial communities to floodplain wetland inundation. *Aquatic Ecology* 43, 4, 843–858. doi: 10.1007/s10452-008-9219-2

Koiter, A.J., Lobb, D.A., Owens, P.N. et al. 2013. Investigating the role of connectivity and scale in assessing the sources of sediment in an agricultural watershed in the Canadian prairies using sediment source fingerprinting. *Journal of Soils and Sediments* 13, 10, 1676–1691. doi: 10.1007/s11368-013-0762-7

Körner, C. and Paulsen, J. 2004. A world-wide study of high altitude treeline tempera-tures. *Journal of Biogeography* 31, 5, 713–732. doi: 10.1111/j.1365-2699.2003.01043.x

Kostaschuk, R., Best, J., Villard, P. et al. 2005. Measuring flow velocity and sediment transport with an acoustic Doppler current profiler. *Geomorphology* 68, 1–2, 25–37. doi: 10.1016/j.geomorph.2004.07.012

Kramer, C. and Gleixner, G. 2008. Soil organic matter in soil depth profiles: Distinct carbon preferences of microbial groups during carbon transformation. *Soil Biology and Biochemistry* 40, 2, 425–433. doi: 10.1016/j.soilbio.2007.09.016

Kristensen, E., Mangion, P., Tang, M. et al. 2011. Microbial carbon oxidation rates and pathways in sediments of two Tanzanian mangrove forests. *Biogeochemistry* 103, 1–3, 143–158. doi: 10.1007/s10533-010-9453-2

Krüger, F., Schwartz, R., Kunert, M. et al. 2006. Methods to calculate sedimentation rates of floodplain soils in the middle region of the Elbe River. *Acta Hydrochimica et Hydrobiologica* 34, 3, 175–187. doi: 10.1002/aheh.200400628

Kuhn, N.J., Hoffmann, T., Schwanghart, W. et al. 2009. Agricultural soil erosion and global carbon cycle: Controversy over? *Earth Surface Processes and Landforms* 34, 7, 1033–1038. doi: 10.1002/esp.1796

Kump, L.R., Brantley, S.L., and Arthur, M.A. 2000. Chemical weathering, atmospheric CO_2, and climate. *Annual Review of Earth and Planetary Sciences* 28, 1, 611–667. doi: 10.1146/annurev.earth.28.1.611

Lagomasino, D., Fatoyinbo, T., Lee, S. et al. 2019. Measuring mangrove carbon loss and gain in deltas. *Environmental Research Letters* 14, 2, 1–10. doi: 10.1088/1748-9326/aaf0de

Laine, A., Sottocornola, M., Kiely, G. et al. 2006. Estimating net ecosystem exchange in a patterned ecosystem: Example from blanket bog. *Agricultural and Forest Meteorology* 138, 1–4, 231–243. doi: 10.1016/j.agrformet.2006.05.005

Lal, R. 1995. Global soil erosion by water and carbon dynamics, in Lal, R., Kimble, J., Levine, E. et al. (eds.), *Soils and Global Change*. CRC Press, London, pp. 131–142.

Lal, R. 2003. Soil erosion and the global carbon budget. *Environment International* 29, 4, 437–450. doi: 10.1016/s0160-4120(02)00192-7

Lal, R. 2004. Soil carbon sequestration impacts on global climate change and food secu-rity. *Science* 304, 5677, 1623–1627.

Lal, R. 2008. Sequestration of atmospheric CO_2 in global carbon pools. *Energy and Environmental Science* 1, 86–100. doi: 10.1039/B809492F

Lambert, C.P. and Walling, D.E. 1987. Floodplain sedimentation: A preliminary inves-tigation of contemporary deposition within the lower reaches of the River Culm, Devon, UK. *Geografiska Annaler Series A Physical Geography* 69, 3–4, 393–404. doi: 10.2307/521353

Langer, U. and Rinklebe, J. 2009. Lipid biomarkers for assessment of microbial commu-nities in floodplain soils of the Elbe River (Germany). *Wetlands* 29, 1, 353–362. doi: 10.1672/08-114.1

Larsen, I.J., Almond, P.C., Eger, A. et al. 2014. Rapid soil production and weathering in the Southern Alps, New Zealand. *Science* 343, 6171, 637–640. doi: 10.1126/science.1244908

Larson, M.G., Booth, D.B., and Morley, S.A. 2001. Effectiveness of large woody debris in stream rehabilitation projects in urban basins. *Ecological Engineering* 18, 2, 211–226. doi: 10.1016/S0925-8574(01)00079-9

Lawler, D.M. 1993. The measurement of river bank erosion and lateral channel change: A review. *Earth Surface Processes and Landforms* 18, 9, 777–821. doi: 10.1002/esp.3290180905

Lee, E.-J., Yoo, G.-Y., Jeong, Y. et al. 2015. Comparison of UV–VIS and FDOM sensors for in situ monitoring of stream DOC concentrations. *Biogeosciences* 12, 10, 3109–3118. doi: 10.5194/bg-12-3109-2015

Lee, H., Schuur, E.A.G., Inglett, K.S. et al. 2012. The rate of permafrost carbon release under aerobic and anaerobic conditions and its potential effects on climate. *Global Change Biology* 18, 2, 515–527. doi: 10.1111/j.1365-2486.2011.02519.x

Lees, K.J., Quaife, T., Artz, R.R.E. et al. 2018. Potential for using remote sensing to estimate carbon fluxes across northern peatlands – A review. *Science of the Total Environment* 615, 857–874. doi: 10.1016/j.scitotenv.2017.09.103

Leithold, E.L., Blair, N.E., and Perkey, D.W. 2006. Geomorphologic controls on the age of particulate organic carbon from small mountainous and upland rivers. *Global Biogeochemical Cycles* 20, 3, 1–11. doi: 10.1029/2005GB002677

Leithold, E.L., Blair, N.E., and Wegmann, K.W. 2016. Source-to-sink sedimentary systems and global carbon burial: A river runs through it. *Earth-Science Reviews* 153, 30–42. doi: 10.1016/j.earscirev.2015.10.011

Lenton, T.M. and Britton, C. 2006. Enhanced carbonate and silicate weathering accelerates recovery from fossil fuel CO_2 perturbations. *Global Biogeochemical Cycles* 20, 3, 1–12. doi: 10.1029/2005GB002678

Leopold, L.B. and Wolman, M.G. 1970. Flood plains, in Dury, G.H. (ed.), *Rivers and River Terraces*. Palgrave Macmillan, London, pp. 166–196.

Leopold, L.B., Wolman, M.G., and Miller, J.P. 1964. *Fluvial Processes in Geomorphology*. Freeman and Company, San Francisco, CA.

Lepori, F., Palm, D., and Malmqvist, B. 2005. Effects of stream restoration on ecosystem functioning: Detritus retentiveness and decomposition. *Journal of Applied Ecology* 42, 2, 228–238. doi: 10.1111/j.1365-2664.2004.00965.x

Lewin, J. and Macklin, M.G. 2003. Preservation potential for Late Quaternary river alluvium. *Journal of Quaternary Science* 18, 2, 107–120. doi: 10.1002/jqs.738

Lewin, J., and Macklin, M.G. 2014. Marking time in Geomorphology: Should we try to formalise an Anthropocene definition? *Earth Surface Processes and Landforms* 39, 1, 133–137. doi: 10.1002/esp.3484

Lewin, J., Macklin, M.G., and Johnstone, E. 2005. Interpreting alluvial archives: Sedimentological factors in the British Holocene fluvial record. *Quaternary Science Reviews* 24, 16–17, 1873–1889. doi: 10.1016/j.quascirev.2005.01.009

Li, C., Grayson, R., Holden, J. et al. 2018. Erosion in peatlands: Recent research progress and future directions. *Earth-Science Reviews* 185, 870–886. doi: 10.1016/j.earscirev.2018.08.005

Li, G. and Elderfield, H. 2013. Evolution of carbon cycle over the ast 100 million years. *Geochimica et Cosmochimica Acta* 103, 11–25. doi: 10.1016/j.gca.2012.10.014

Li, M., Peng, C., Wang, M. et al. 2017a. The carbon flux of global rivers: A re-evaluation of amount and spatial patterns. *Ecological Indicators* 80, 40–51. doi: 10.1016/j. ecolind.2017.04.049

Li, P., Holden, J., Irvine, B. et al. 2017b. Erosion of Northern Hemisphere blanket peatlands under 21st-century climate change. *Geophysical Research Letters* 44, 8, 3615–3623. doi: 10.1002/2017GL072590

Li, Y., Ruan, H., Zou, X. et al. 2005. Response of major soil decomposers to landslide disturbance in a Puerto Rican rainforest. *Soil Science* 170, 3, 202–211. doi: 10.1097/00010694-200503000-00006

Lim, M., Dunning, S.A., Burke, M. et al. 2015. Quantification and implications of change in organic carbon bearing coastal dune cliffs: A multiscale analysis from the Northumberland coast, UK. *Remote Sensing of Environment* 163, 1–12. doi: 10.1016/j. rse.2015.01.034

Limpens, J., Berendse, F., Blodau, C. et al. 2008. Peatlands and the carbon cycle: From local processes to global implications – A synthesis. *Biogeosciences* 5, 5, 1475–1491. doi: 10.5194/bg-5-1475-2008

Lin, G.-W., Chen, H., Chen, Y.-H. et al. 2008. Influence of typhoons and earthquakes on rainfall-induced landslides and suspended sediments discharge. *Engineering Geology* 97, 1–2, 32–41. doi: 10.1016/j.enggeo.2007.12.001

Lindsay, R., Charman, D., Everingham, F. et al. 1988. *The Flow Country: The Peatlands of Caithness and Sutherland*. Joint Nature Conservation Committee, Peterborough.

Linn, D.M. and Doran, J.W. 1984. Effect of water-filled pore space on carbon dioxide and nitrous oxide production in tilled and nontilled soils 1. *Soil Society of America Journal* 48, 6, 1267–1272. doi: 10.2136/sssaj1984.03615995004800060013x

Lipson, D.A., Schmidt, S.K., and Monson, R.K. 2000. Carbon availability and temperature control the post-snowmelt decline in alpine soil microbial biomass. *Soil Biology and Biochemistry* 32, 4, 441–448. doi: 10.1016/S0038-0717(99)00068-1

Liu, W.-T. and Zhu, L. 2005. Environmental microbiology-on-a-chip and its future impacts. *Trends Biotechnology* 23, 4, 174–179. doi: 10.1016/j.tibtech.2005.02.004

Liu, S., Bliss, N., Sundquist, E. et al. 2003. Modeling carbon dynamics in vegetation and soil under the impact of soil erosion and deposition. *Global Biogeochemical Cycles* 17, 2, 1–43. https://doi.org/10.1029/2002GB002010

Lloyd, J. and Taylor, J.A. 1994. On the temperature dependence of soil respiration. *Functional Ecology* 8, 3, 315–323. doi: 10.2307/2389824

Lovelock, C.E., Cahoon, D.R., Friess, D.A. et al. 2015. The vulnerability of Indo-Pacific mangrove forests to sea-level rise. *Nature* 526, 559–563. doi: 10.1038/nature15538

Lu, H., Moran, C.J., and Sivapalan, M. 2005. A theoretical exploration of catchment-scale sediment delivery. *Water Resources Research* 41, 9, 1–15. doi: 10.1029/2005WR004018

Ludwig, W., Amiotte-Suchet, P., and Probst, J.-L. 1996a. River discharges of carbon to the world's oceans: Determining local inputs of alkalinity and of dissolved and particulate organic carbon. *Comptes Rendus de l'Academie des Sciences* 323, 1007–1014. https://oatao.univ-toulouse.fr/3498/

Ludwig, W., Probst, J.-L., and Kempe, S. 1996b. Predicting the oceanic input of organic carbon by continental erosion. *Global Biogeochemical Cycles* 10, 1, 23–41. doi: 10.1029/95GB02925

Luo, T., Pan, Y., Ouyang, H. et al. 2004. Leaf area index and net primary productivity along subtropical to alpine gradients in the Tibetan Plateau. *Global Ecology and Biogeography* 13, 4, 345–358. doi: 10.1111/j.1466-822X.2004.00094.x

Lupker, M., France-Lanord, C., Galy, V. et al. 2012. Predominant floodplain over mountain weathering of Himalayan sediments (Ganga Basin). *Geochimica et Cosmochimica Acta* 84, 410–432. doi: 10.1016/j.gca.2012.02.001

Lupker, M., France-Lanord, C., Galy, V. et al. 2013. Increasing chemical weathering in the Himalayan system since the Last Glacial Maximum. *Earth and Planetary Science Letters* 365, 243–252. doi: 10.1016/j.epsl.2013.01.038

Lützow, M.V., Kögel-Knabner, I., Ekschmitt, K. et al. 2006. Stabilization of organic matter in temperate soils: Mechanisms and their relevance under different soil conditions – A review. *European Journal of Soil Science* 57, 4, 426–445. doi: 10.1111/j.1365-2389.2006.00809.x

MacDonell, S. and Fitzsimons, S. 2008. The formation and hydrological significance of cryoconite holes. *Progress in Physical Geography* 32, 6, 595–610. doi: 10.1177/0309133308101382

Macklin, M.G., and Lewin, J. 2008. Alluvial responses to the changing Earth system. *Earth Surface Processes and Landforms* 33, 9, 1374–1395. doi: 10.1002/esp.1714

Macreadie, P.I., Hughes, A.R., and Kimbro, D.L. 2013. Loss of 'Blue Carbon' from coastal salt marshes following habitat disturbance. *PLOS ONE* 8, 7, 1–8. doi: 10.1371/journal.pone.0069244

Magilligan, F.J. 1985. Historical floodplain sedimentation in the Galena River basin, Wisconsin and Illinois. *Annals of the American Association of Geographers* 75, 4, 583–594. doi: 10.1111/j.1467-8306.1985.tb00095.x

Magilligan, F.J., Graber, B.E., Nislow, K.H. et al. 2016. River restoration by dam removal: Enhancing connectivity at watershed scales. *Elementa: Science of the Anthropocene* 4, 1–14. doi: 10.12952/journal.elementa.000108

Maher, D.T., Santos, I.R., Golsby-Smith, L. et al. 2013. Groundwater-derived dissolved inorganic and organic carbon exports from a mangrove tidal creek: The missing mangrove carbon sink? *Limnology and Oceanography* 58, 2, 475–488. doi: 10.4319/lo.2013.58.2.0475

Maher, K. and Chamberlain, C.P. 2014. Hydrologic regulation of chemical weathering and the geologic carbon cycle. *Science* 343, 6178, 1502–1504. doi: 10.1126/science.1250770

Mangiarotti, S., Martinez, J.-M., Bonnet, M.-P. et al. 2013. Discharge and suspended sediment flux estimated along the mainstream of the Amazon and the Madeira rivers (from *in situ* and MODIS Satellite Data). *International Journal of Applied Earth Observation and Geoinformation* 21, 341–355. doi: 10.1016/j.jag.2012.07.015

Marani, M., D'Alpaos, A., Lanzoni, S. et al. 2011. Understanding and predicting wave erosion of marsh edges. *Geophysical Research Letters* 38, 21, 1–5. doi: 10.1029/2011GL048995

Marschner, B., Brodowski, S., Dreves, A. et al. 2008. How relevant is recalcitrance for the stabilization of organic matter in soils? *Journal of Plant Nutrition and Soil Science* 171, 1, 91–110. doi: 10.1002/jpln.200700049

Mateo, M.A., Cebrián, J., Dunton, K. et al. 2006. Carbon flux in seagrass ecosystems, in Larkum, A.W.D., Orth, R.J., and Duarte, C.M. (eds.), *Seagrasses: Biology, Ecology and Conservation*. Springer, Dordrecht, The Netherlands, pp. 159–192.

McCarty, G.W. and Ritchie, J.C. 2002. Impact of soil movement on carbon sequestration in agricultural ecosystems. *Environmental Pollution* 116, 3, 423–430. doi: 1010.1016/s0269-7491(01)00219-6

McClain, M.E., Boyer, E.W., Dent, C.L. et al. 2003. Biogeochemical hot spots and hot moments at the interface of terrestrial and aquatic ecosystems. *Ecosystems* 6, 301–312. doi: 10.1007/s10021-003-0161-9

McClelland, J.W., Holmes, R.M., Peterson, B.J. et al. 2016. Particulate organic carbon and nitrogen export from major Arctic rivers. *Global Biogeochemical Cycles* 30, 5, 629–643. doi: 10.1002/2015GB005351c

McDowell, W.H. and Asbury, C.E. 1994. Export of carbon, nitrogen, and major ions from three tropical montane watersheds. *Limnology and Oceanography* 39, 1, 111–125. doi: 10.4319/lo.1994.39.1.0111

McGill, W.B. 1996. Review and classification of ten soil organic matter (SOM) models, in Powlson, D.S., Smith, P., and Smith, J.U. (eds.), *Evaluation of Soil Organic Matter Models*, NATO ASI Series. Springer, Berlin and Heidelberg, pp. 111–132. doi: 10.1007/978-3-642-61094-3_9

McIntyre, R.E.S., Adams, M.A., Ford, D.J. et al. 2009. Rewetting and litter addition influence mineralisation and microbial communities in soils from a semi-arid intermittent stream. *Soil Biology and Biochemistry* 41, 1, 92–101. doi: 10.1016/j.soilbio.2008.09.021

McLauchlan, K. 2006. The nature and longevity of agricultural impacts on soil carbon and nutrients: A review. *Ecosystems* 9, 8, 1364–1382. doi: 10.1007/s10021-005-0135-1

McLauchlan, K.K., Hobbie, S.E., and Post, W.M. 2006. Conversion from agriculture to grassland builds soil organic matter on decadal timescales. *Ecological Applications* 16, 1, 143–153. doi: 10.1890/04-1650

McLaughlin, M.J., Fillery, I.R., and Till, A.R. 1992. Operation of the phosphorus, sulphur and nitrogen cycles, in Gifford, R.M. and Barson, M.M. (eds.), *Australia's Renewable Resources: Sustainability and Global Change*, Bureau of Rural Resources Proceedings, Proceedings of the Australian-IGBP Planning Workshop. Canberra, Australia, 14, pp. 67–116.

Mcleod, E., Chmura, G.L., Bouillon, S. et al. 2011. A blueprint for blue carbon: Toward an improved understanding of the role of vegetated coastal habitats in sequestering CO_2. *Frontiers in Ecology and the Environment* 9, 10, 552–560. doi: 10.1890/110004

Meade, R.H. 1994. Suspended sediments of the modern Amazon and Orinoco rivers. *Quaternary International* 21, 29–39. doi: 10.1016/1040-6182(94)90019-1

Mekonnen, M.M. and Hoekstra, A.Y. 2015. Global gray water footprint and water pollution levels related to anthropogenic nitrogen loads to fresh water. *Environmental Science and Technology* 49, 21, 12860–12868. doi: 10.1021/acs.est.5b03191

Merriam, G. 1984. Connectivity: A fundamental ecological characteristic of landscape pattern, in Brandt, J. and Agger, P.A. (eds.), *Proceedings of the First International Seminar on Methodology in Landscape Ecological Research and Planning Rosskilde University Centre Rosskilde*. Roskilde University Centre, Roskilde, Denmark, pp. 5–15.

Mertes, L.A.K. 1994. Rates of flood-plain sedimentation on the central Amazon River. *Geology* 22, 2, 171–174. doi: 10.1130/0091-7613(1994)022<0171:ROFPSO>2.3.CO;2

Mestdagh, I., Sleutel, S., Lootens, P. et al. 2005. Soil organic carbon stocks in verges and urban areas of Flanders, Belgium. *Grass and Forage Science* 60, 2, 151–156. doi: 10.1111/j.1365-2494.2005.00462.x

Meybeck, M. 1982. Carbon, nitrogen, and phosphorus transport by world rivers. *American Journal of Science* 282, 401–450. doi: 10.2475/ajs.282.4.401

Meybeck, M. 1993. Riverine transport of atmospheric carbon: Sources, global typology and budget. *Water, Air and Soil Pollution* 70, 1–4, 443–463. doi: 10.1007/bf01105015

Meybeck, M. 2003. Global analysis of river systems: From Earth system controls to Anthropocene syndromes. *Philosophical Transactions of the Royal Society B: Biological Sciences* 358, 1440, 1935–1955. doi: 10.1098/rstb.2003.1379

Meybeck, M. and Helmer, R. 1989. The quality of rivers: From pristine stage to global pollution. *Global and Planetary Change* 1, 4, 283–309. doi: 10.1016/0921-8181(89)90007-6

Middelkoop, H. 2002. Reconstructing floodplain sedimentation rates from heavy metal profiles by inverse modelling. *Hydrological Processes* 16, 1, 47–64. doi: 10.1002/hyp.283

Middelkoop, H. and Asselman, N.E.M. 1998. Spatial variability of flood-plain sedimentation at the event scale in the Rhine–Meuse delta, The Netherlands. *Earth Surface Processes and Landforms* 23, 6, 561–573. doi: 10.1002/(SICI)1096-9837(199806)23:6<561::AID-ESP870>3.0.CO;2-5

Millennium Ecosystem Assessment 2005. *Ecosystems and Human Well-Being: Synthesis*. Island Press, Washington, DC.

Millot, R., Gaillardet, J., Dupré, B. et al. 2002. The global control of silicate weathering rates and the coupling with physical erosion: New insights from rivers of the Canadian shield. *Earth and Planetary Science Letters* 196, 1–2, 83–98. doi: 10.1016/S0012-821X(01)00599-4

Mildowski, D.T., Mudd, S.M., and Mitchard, E.T.A. 2015. Erosion rates as a potential bottom-up control of forest structural characteristics in the Sierra Nevada Mountains. *Ecology* 96, 31–38.

Mitchell, E.A.T., Gilbert, D., Buttler, A. et al. 2003. Structure of microbial communities in Sphagnum peatlands and effect of atmospheric carbon dioxide enrichment. *Microbial Ecology* 46, 2, 187–199. doi: 10.1007/BF03036882

Mitsch, W.J., Dorage, C.L., and Wiemhoff, J.R. 1977. Forested wetlands for water resource management in southern Illinois. Research Report No. 132 Illinois Water Resources Center: Urbana, Illinois.

Mitsch, W.J., Dorage, C.L., and Wiemhoff, J.R. 1979. Ecosystem dynamics and a phosphorus budget of an alluvial cypress swamp in southern Illinois. *Ecology* 60, 6, 1116–1124 https://doi.org/10.2307/1936959.

Montgomery, D.R. 2007. Soil erosion and agricultural sustainability. *Proceedings of the National Academy of Sciences of the United States of America* 104, 33, 13268–13272. doi: 10.1073/pnas.0611508104

Moody, C.S., Worrall, F., Clay, G.D. et al. 2018. A molecular budget for a peatland based upon ^{13}C solid-state nuclear magnetic resonance. *Journal of Geophysical Research: Biogeosciences* 123, 2, 547–560. doi: 10.1002/2017JG004312

Moody, C.S., Worrall, F., Evans, C.D. et al. 2013. The rate of loss of dissolved organic carbon (DOC) through a catchment. *Journal of Hydrology* 492, 139–150. doi: 10.1016/j.jhydrol.2013.03.016

Moore, S., Evans, C.D., Page, S.E. et al. 2013. Deep instability of deforested tropical peatlands revealed by fluvial organic carbon fluxes. *Nature* 493, 660–663. doi: 10.1038/nature11818

Moore, T.R. and Knowles, R. 1989. The influence of water table levels on methane and carbon dioxide emissions from peatland soils. *Canadian Journal of Soil Science* 69, 1, 33–38. doi: 10.4141/cjss89-004

Moran, M.A. and Hodson, R.E. 1989. Formation and bacterial utilization of dissolved organic carbon derived from detrital lignocellulose. *Limnology and Oceanography* 34, 6, 1034–1047. doi: 10.4319/lo.1989.34.6.1034

Moran, M.A., Sheldon, W.M., and Sheldon, J.E. 1999. Biodegradation of riverine dissolved organic carbon in five estuaries of the southeastern United States. *Estuaries* 22, 1, 55–64. doi: 10.2307/1352927

Moran, M.A., Sheldon, W.M., and Zepp, R.G. 2000. Carbon loss and optical property changes during long-term photochemical and biological degradation of estuarine dissolved organic matter. *Limnology and Oceanography* 45, 6, 1254–1264. doi: 10.4319/lo.2000.45.6.1254

Moreira-Turcq, P., Jouanneau, J.M., Turcq, B. et al. 2004. Carbon sedimentation at Lago Grande de Curuai, a floodplain lake in the low Amazon region: Insights into sedimentation rates. *Palaeogeography, Palaeoclimatology, Palaeoecology* 214, 1–2, 27–40. doi: 10.1016/j.palaeo.2004.06.013

Morgan, R.P.C. 2009. *Soil Erosion and Conservation*. John Wiley & Sons, Oxford, UK.

Morozova, G.S. and Smith, N.D. 2003. Organic matter deposition in the Saskatchewan River floodplain (Cumberland Marshes, Canada): Effects of progradational avulsions. *Sedimentary Geology* 157, 1–2, 15–29. doi: 10.1016/S0037-0738(02)00192-6

Mudd, S.M.,and Fagherazzi, S. 2016. Salt marsh ecosystems: Tidal flow, vegetation and carbon dynamics, in Johnson, E.A. and Martin, Y.E. (eds.), *A Biogeoscience Approach to Ecosystems*. Cambridge University Press, Cambridge, UK, pp. 407–434.

Myers, R.T., Zak, D.R., White, D.C. et al. 2001. Landscape-level patterns of microbial community composition and substrate use in upland forest ecosystems. *Soil Society of America Journal* 65, 2, 359–367. doi: 10.2136/sssaj2001.652359x

Myhre, G., Shindell, D., Bréon, F.-M. et al. 2013. Anthropogenic and natural radiative forcing, in Stocker, T.F., Qin, D., Plattner, G.-K. et al. (eds.), *IPCC 2013: Climate Change 2013: The Physical Science Basis. Contribution of Working Group I to the Fifth Assessment Report of the Intergovernmental Panel on Climate Change.* Cambridge University Press, Cambridge, UK, pp. 659–740.

Nanson, G.C. and Croke, J.C. 1992. A genetic classification of floodplains. *Geomorphology* 4, 6, 459–486. doi: 10.1016/0169-555X(92)90039-Q

NASA 2016. http://earthobservatory.nasa.gov/Features/CarbonCycle/page2.php (accessed 8 February 2017)

Naylor, L.A., Viles, H.A., and Carter, N.E.A. 2002. Biogeomorphology revisited: Looking towards the future. *Geomorphology* 47, 1, 3–14. doi: 10.1016/S0169-555X(02)00137-X

Newcomer, T.A., Kaushal, S.S., Mayer, P.M. et al. 2012. Influence of natural and novel organic carbon sources on denitrification in forest, degraded urban, and restored streams. *Ecological Monographs* 82, 4, 449–466. doi: 10.1890/12-0458.1

Nicholas, A.P. and Walling, D.E. 1996. The significance of particle aggregation in the overbank deposition of suspended sediment on river floodplains. *Journal of Hydrology* 186, 1–4, 275–293. doi: 10.1016/S0022-1694(96)03023-5

Nicoll, T.J. and Hickin, E.J. 2010. Planform geometry and channel migration of confined meandering rivers on the Canadian prairies. *Geomorphology* 116, 1–2, 37–47. doi: 10.1016/j.geomorph.2009.10.005

Nilsson, C., Jansson, R., Malmqvist, B. et al. 2007. Restoring riverine landscapes: The challenge of identifying priorities, reference states, and techniques. *Ecology and Society* 12, 1, 1–7. doi: 10.5751/ES-02030-120116

Noe, G.B. and Hupp, C.R. 2005. Carbon, nitrogen, and phosphorus accumulation in floodplains of Atlantic coastal plain rivers, USA. *Ecological Applications* 15, 4, 1178–1190. doi: 10.1890/04-1677

Notebaert, B. and Piégay, H. 2013. Multi-scale factors controlling the pattern of floodplain width at a network scale: The case of the Rhône basin, France. *Geomorphology* 200, 155–171. doi: 10.1016/j.geomorph.2013.03.014

Obu, J., Lantuit, H., Grosse, G. et al. 2017. Coastal erosion and mass wasting along the Canadian Beaufort Sea based on annual airborne LiDAR elevation data. *Geomorphology* 293, 331–346. doi: 10.1016/j.geomorph.2016.02.014

O'Connell, M., Baldwin, D.S., Robertson, A.I. et al. 2000. Release and bioavailability of dissolved organic matter from floodplain litter: Influence of origin and oxygen levels. *Freshwater Biology* 45, 3, 333–342. doi: 10.1111/j.1365-2427.2000.00627.x

Opperman, J., Luster, R., McKenney, B. et al. 2010. Ecologically functional floodplains: Connectivity, flow regime, and scale. *Journal of the American Water Resources Association* 46, 2, 211–226. doi: 10.1111/j.1752-1688.2010.00426.x

Ouyang, X. and Lee, S.Y. 2014. Updated estimates of carbon accumulation rates in coastal marsh sediments. *Biogeosciences* 11, 18, 5057–5071. doi: 10.5194/bg-11-5057-2014

Ouyang, X., Lee, S.Y., and Connolly, R.M. 2017. The role of root decomposition in global mangrove and saltmarsh carbon budgets. *Earth-Science Reviews* 166, 53–63. doi: 10.1016/j.earscirev.2017.01.004

Owens, P.N., Walling, D.E., and Leeks, G.J.L. 1999. Deposition and storage of fine-grained sediment within the main channel system of the River Tweed, Scotland. *Earth Surface Processes and Landforms* 24, 12, 1061–1076. doi: 10.1002/(SICI)1096-9837(199911)24:12<1061::AID-ESP35>3.0.CO;2-Y

Page, M., Trustrum, N., Brackley, H. et al. 2004. Erosion-related soil carbon fluxes in a pastoral steepland catchment, New Zealand. *Agriculture, Ecosystems and Environment* 103, 3, 561–579. doi: 10.1016/j.agee.2003.11.010

Palmer, M.A., Menninger, H.L., and Bernhardt, E. 2010. River restoration, habitat heterogeneity and biodiversity: A failure of theory or practice? *Freshwater Biology* 55, 205–222. doi: 10.1111/j.1365-2427.2009.02372.x

Parsons, A.J., Wainwright, J., Brazier, R.E. et al. 2006. Is sediment delivery a fallacy? *Earth Surface Processes and Landforms* 31, 10, 1325–1328. doi: 10.1002/esp.1395

Parton, W.J. 1996. The CENTURY model, in Powlson, D.S., Smith, P., and Smith, J.U. (eds.), *Evaluation of Soil Organic Matter Models*. Springer, Berlin, Germany, pp. 283–291.

Parton, W.J., Schimel, D.S., Cole, C.V. et al. 1987. Analysis of factors controlling soil organic matter levels in Great Plains grasslands. *Soil Science Society of America Journal* 51, 5, 1173–1179. doi: 10.2136/sssaj1987.03615995005100050015x

Pawlik, Ł. 2013. The role of trees in the geomorphic system of forested hillslopes – A review. *Earth-Science Reviews* 126, 250–265. doi: 10.1016/j.earscirev.2013.08.007

Pawson, R.R., Evans, M.G., and Allott, T.E.H.A. 2012. Fluvial carbon flux from headwater peatland streams: Significance of particulate carbon flux. *Earth Surface Processes and Landforms* 37, 11, 1203–1212. doi: 10.1002/esp.3257

Pawson, R.R., Lord, D.R., Evans, M.G. et al. 2008. Fluvial organic carbon flux from an eroding peatland catchment, southern Pennines, UK. *Hydrology and Earth System Sciences* 12, 625–634. doi: 10.5194/hess-12-625-2008

Pei, T., Qin, C.-Z., Zhu, A.-X. et al. 2010. Mapping soil organic matter using the topographic wetness index: A comparative study based on different flow-direction algorithms and kriging methods. *Ecological Indicators* 10, 3, 610–619. doi: 10.1016/j.ecolind.2009.10.005

Pemberton, M. 2005. Australian peatlands: A brief consideration of their origin, distribution, natural values and threats. *Journal of the Royal Society of Western Australia* 88, 81–89. https://www.rswa.org.au/publications/Journal/88(3)/volume88part381-89.pdf

Penman, D.E., Caves Rugenstein, J.K., Ibarra, D.E. et al. 2020. Silicate weathering as a feedback and forcing in Earth's climate and carbon cycle. *Earth-Science Reviews* 209, 103298. doi: 10.1016/j.earscirev.2020.103298

Pethick, J.S. 1992. Saltmarsh geomorphology, in Allen, J.R.L. and Pye, K. (eds.), *Saltmarshes: Morphodynamics, Conservation and Engineering Significance*. Cambridge University Press, Cambridge, UK, pp. 41–62.

Petsch, S.T. 2014. Weathering of organic carbon, in Holland, H.D. and Turekian, K.K. (eds.), *Treatise on Geochemistry*, 2nd edn. Elsevier. https://doi.org/10.1016/B978-0-08-095975-7.01013-5.

Pfister, L., McDonnell, J.J., Wrede, S. et al. 2009. The rivers are alive: On the potential for diatoms as a tracer of water source and hydrological connectivity. *Hydrological Processes* 23, 19, 2841–2845. doi: 10.1002/hyp.7426

Phillips, J.D. 1995. Biogeomorphology and landscape evolution: The problem of scale. *Geomorphology* 13, 1–4, 337–347. doi: 10.1016/0169-555X(95)00023-X

Phillips, J.D., Marden, M., and Gomez, B. 2007. Residence time of alluvium in an aggrading fluvial system. *Earth Surface Processes and Landforms* 32, 2, 307–316. doi: 10.1002/esp.1385

Phillips, J.D., Slattery, M.C., and Musselman, Z.A. 2004. Dam-to-delta sediment inputs and storage in the lower trinity river, Texas. *Geomorphology* 62, 1–2, 17–34. doi: 10.1016/j.geomorph.2004.02.004

Pithan, F. and Mauritsen, T. 2014. Arctic amplification dominated by temperature feedbacks in contemporary climate models. *Nature Geoscience* 7, 181–184. doi: 10.1038/ngeo2071

Pollock, M.M., Beechie, T.J., Wheaton, J.M. et al. 2014. Using beaver dams to restore incised stream ecosystems. *BioScience* 64, 4, 279–290. doi: 10.1093/biosci/biu036

Polvi, L.E., Nilsson, C., and Hasselquist, E.M. 2014. Potential and actual geomorphic complexity of restored headwater streams in northern Sweden. *Geomorphology* 210, 98–118. doi: 10.1016/j.geomorph.2013.12.025

Polvi, L.E. and Wohl, E. 2013. Biotic drivers of stream planform: Implications for understanding the past and restoring the future. *BioScience* 63, 6, 439–452. doi: 10.1525/bio.2013.63.6.6

Pope, G.A., Meierding, T.C., and Paradise, T.R. 2002. Geomorphology's role in the study of weathering of cultural stone. *Geomorphology* 47, 2–4, 211–225. doi: 10.1016/S0169-555X(02)00098-3

Porder, S., Johnson, A.H., Xing, H.X. et al. 2015. Linking geomorphology, weathering and cation availability in the Luquillo Mountains of Puerto Rico. *Geoderma* 249–250, 100–110. doi: 10.1016/j.geoderma.2015.03.002

Porder, S., Vitousek, P.M., Chadwick, O.A. et al. 2007. Uplift, erosion, and phosphorus limitation in terrestrial ecosystems. *Ecosystems* 10, 159–171. doi: 10.1007/s10021-006-9011-x

Pouyat, R., Groffman, P., Yesilonis, I. et al. 2002. Soil carbon pools and fluxes in urban ecosystems. *Environmental Pollution* 116, 107–118. doi: 10.1016/S0269-7491(01)00263-9

Pouyat, R.V., Yesilonis, I.D., and Nowak, D.J. 2006. Carbon storage by urban soils in the United States. *Journal of Environmental Quality* 35, 4, 1566–1575. doi: 10.2134/jeq2005.0215

Price, S.J., Ford, J.R., Cooper, A.H. et al. 2011. Humans as major geological and geomorphological agents in the Anthropocene: The significance of artificial ground in Great Britain. *Philosophical Transactions of the Royal Society A: Mathematical Physical and Engineering Sciences* 369, 1938, 1056–1084. doi: 10.1098/rsta.2010.0296

Quinton, J.N., Govers, G., Van Oost, K. et al. 2010. The impact of agricultural soil erosion on biogeochemical cycling. *Nature Geoscience* 3, 311–314. doi: 10.1038/ngeo838

Raich, J.W., Russell, A.E., and Vitousek, P.M. 1997. Primary productivity and ecosystem development along and elevational gradient on Mauna Loa, Hawai'i. *Ecology* 78, 3, 707–721. doi: 10.1890/0012-9658(1997)078[0707:PPAEDA]2.0.CO;2

Raich, J.W. and Schlesinger, W.H. 1992. The global carbon dioxide flux in soil respiration and its relationship to vegetation and climate. *Tellus B* 44, 81–99. doi: 10.1034/j.1600-0889.1992.t01-1-00001.x

Raich, J.W. and Tufekciogul, A. 2000. Vegetation and soil respiration: Correlations and controls. *Biogeochemistry* 48, 1, 71–90. doi: 10.1023/A: 1006112000616

Rainato, R., Mao, L., García-Rama, A. et al. 2017. Three decades of monitoring in the Rio Cordon instrumented basin: Sediment budget and temporal trend of sediment yield. *Geomorphology* 291, 45–56. doi: 10.1016/j.geomorph.2016.03.012

Ramchunder, S.J., Brown, L.E., and Holden, J. 2009. Environmental effects of drainage, drain-blocking and prescribed vegetation burning in UK upland peatlands. *Progress in Physical Geography: Earth and Environment* 33, 1, 49–79. doi: 10.1177/0309133309105245

Ramos Scharrón, C.E., Castellanos, E.J. et al. 2012. The transfer of modern organic carbon by landslide activity in tropical montane ecosystems: Effects on landslides on carbon budgets. *Journal of Geophysical Research: Biogeosciences* 117, G3, 1–18. doi: 10.1029/2011JG001838

Ran, L., Lu, X.X., and Xin, Z. 2014. Erosion-induced massive organic carbon burial and carbon emission in the Yellow River basin, China. *Biogeosciences* 11, 945–959. doi: 10.5194/bg-11-945-2014

Randle, T.J., Bountry, J.A., Ritchie, A. et al. 2015. Large-scale dam removal on the Elwha River, Washington, USA: Erosion of reservoir sediment. *Geomorphology* 246, 709–728. doi: 10.1016/j.geomorph.2014.12.045

Rawlins, B.G., Harris, J., Price, S. et al. 2015. A review of climate change impacts on urban soil functions with examples and policy insights from England, UK. *Soil Use and Management* 31, 46–61. doi: 10.1111/sum.12079

Raymo, M.E. and Ruddiman, W.F. 1992. Tectonic forcing of late Cenozoic climate. *Nature* 359, 117–122. doi: 10.1038/359117a0

Reckendorfer, W., Funk, A., Gschöpf, C. et al. 2013. Aquatic ecosystem functions of an isolated floodplain and their implications for flood retention and management. *Journal of Applied Ecology* 50, 1, 119–128. doi: 10.1111/1365-2664.12029

Regnier, P., Friedlingstein, P., Ciais, P. et al. 2013. Anthropogenic perturbation of the carbon fluxes from land to ocean. *Nature Geoscience* 6, 597–607. doi: 10.1038/ngeo1830

Reid, L.M. and Page, M.J. 2003. Magnitude and frequency of landsliding in a large New Zealand catchment. *Geomorphology* 49, 1–2, 71–88. doi: 10.1016/S0169-555X(02)00164-2

Restrepo, C., Walker, L.R., Shiels, A.B. et al. 2009. Landsliding and its multiscale influence on mountainscapes. *BioScience* 59, 8, 685–698. http://www.bioone.org/doi/full/10.1525/bio.2009.59.8.10

Rice, C.M. 1949. *Dictionary of Geological Terms*. Brothers, Ann Arbor, Michigan, USA.

Rice, S. 1998. Which tributaries disrupt downstream fining along gravel-bed rivers? *Geomorphology* 22, 1, 39–56. doi: 10.1016/S0169-555X(97)00052-4

Rice, S.P., Ferguson, R.I., and Hoey, T.B. 2006. Tributary control of physical hetero-geneity and biological diversity at river confluences. *Canadian Journal of Fisheries and Aquatic Sciences* 63, 11, 2553–2566. doi: 10.1139/f06-145

Ricker, M.C., Donohue, S.W., Stolt, M.H. et al. 2012. Development and application of multi-proxy indices of land use change for riparian soils in southern New England, USA. *Ecological Applications* 22, 2, 487–501. doi: 10.1890/11-1640.1

Ridgwell, A. and Zeebe, R.E. 2005. The role of the global carbonate cycle in the reg-ulation and evolution of the Earth system. *Earth and Planetary Science Letters* 234, 299–315. doi: 10.1016/j.epsl.2005.03.006

Ridgwell, A.J. 2002. Dust in the Earth system: The biogeochemical linking of land, air and sea. *Philosophical Transactions of the Royal Society A: Mathematical Physical and Engineering Sciences* 360, 1801, 2905–2924. doi: 10.1098/rsta.2002.1096

Riggsbee, J.A., Julian, J.P., Doyle, M.W. et al. 2007. Suspended sediment, dissolved organic carbon, and dissolved nitrogen export during the dam removal process. *Water Resources Research* 43, 9, 1–16. doi: 10.1029/2006WR005318

Rinklebe, J. and Langer, U. 2006. Microbial diversity in three floodplain soils at the Elbe River (Germany). *Soil Biology and Biochemistry* 38, 8, 2144–2151. doi: 10.1016/j.soilbio.2006.01.018

Rinnan, R. and Bååth, E. 2009. Differential utilization of carbon substrates by bacteria and fungi in tundra soil. *Applied and Environmental Microbiology* 75, 11, 3611–3620. doi: 10.1128/AEM.02865-08

Rochefort, L., Quinty, F., Campeau, S. et al. 2003. North American approach to the restoration of *Sphagnum* dominated peatlands. *Wetlands Ecology and Management* 11, 3–20. doi: 10.1023/a: 1022011027946

Rogers, R.D. and Schumm, S.A. 1991. The effect of sparse vegetative cover on erosion and sediment yield. *Journal of Hydrology* 123, 1–2, 19–24. doi: 10.1016/0022-1694(91)90065-P

Rooney-Varga, J.N., Giewat, M.W., Duddleston, K.N. et al. 2007. Links between archaeal community structure, vegetation type and methanogenic pathway in Alaskan peatlands. *FEMS Microbiology Ecology* 60, 20, 240–251. doi: 10.1111/j.1574-6941.2007.00278.x

Rowson, J.G., Worrall, F., and Evans, M.G. 2013. Predicting soil respiration from peat-lands. *Science of the Total Environment* 442, 397–404. doi: 10.1016/j.scitotenv.2012.10.021

Ruddiman, W.F. 2008. *Earth's Climate Past and Future*, 2nd edn. Freeman, New York.

Running, S.W. 2008. Ecosystem disturbance, carbon, and climate. *Science* 321, 5889, 652–653. doi: 10.1126/science.1159607

Rustad, L.E., Huntington, T.G., and Boone, R.D. 2000. Controls on soil respira-tion: Implications for climate change. *Biogeochemistry* 48, 1, 1–6. doi: 10.1023/A: 1006255431298

Sagova-Mareckova, M., Zadorova, T., Penizek, V. et al. 2016. The structure of bacte-rial communities along two vertical profiles of a deep colluvial soil. *Soil Biology and Biochemistry* 101, 65–73. doi: 10.1016/j.soilbio.2016.06.026

Sahoo, K. and Dhal, N.K. 2009. Potential microbial diversity in mangrove ecosystems: A review. *International Journal of Molecular Sciences* 38, 2, 249–256.

Saintilan, N., Rogers, K., Mazumder, D. et al. 2013. Allochthonous and autochthonous contributions to carbon accumulation and carbon store in southeastern Australian coastal wetlands. *Estuarine, Coastal and Shelf Science* 128, 84–92. doi: 10.1016/j.ecss.2013.05.010

Samaritani, E., Shrestha, J., Fournier, B. et al. 2011. Heterogeneity of soil carbon pools and fluxes in a channelized and a restored floodplain section (Thur River, Switzerland). *Hydrology and Earth System Sciences* 15, 1757–1769. doi: 10.5194/hess-15-1757-2011

Sayer, E.J., Oliver, A.E., Fridley, J.D. et al. 2017. Links between soil microbial communities and plant traits in a species-rich grassland under long-term climate change. *Ecology and Evolution* 7, 3, 855–862. doi: 10.1002/ece3.2700

Scharlemann, J.P.W., Tanner, E.V.J., Hiederer, R. et al. (2014). Global soil carbon: understanding and managing the largest terrestrial carbon pool. *Carbon Management* 5, 1, 81–91. https://doi.org/10.4155/cmt.13.77

Schimel, D.S., VEMAP Participants, and Braswell, B.H. 1997. Continental scale variability in ecosystem processes: Models, data, and the role of disturbance. *Ecological Monographs* 67, 2, 251–271. doi: 10.1890/0012-9615(1997)067[0251:CSVIEP]2.0.CO;2

Schlesinger, W.H. and Melack, J.M. 1981. Transport of organic carbon in the world's rivers. *Tellus* 33, 2, 172–187. doi: 10.3402/tellusa.v33i2.10706

Schlünz, B. and Schneider, R.R. 2000. Transport of terrestrial organic carbon to the oceans by rivers: Re-estimating flux- and burial rates. *International Journal of Earth Sciences* 88, 4, 599–606. doi: 10.1007/s005310050290

Schmidt, J. and Hewitt, A. 2004. Fuzzy land element classification from DTMs based on geometry and terrain position. *Geoderma* 121, 3–4, 243–256. doi: 10.1016/j.geoderma.2003.10.008

Schmidt, M.W.I., Torn, M.S., Abiven, S. et al. 2011. Persistence of soil organic matter as an ecosystem property. *Nature* 478, 49–56. doi: 10.1038/nature10386

Schmidt, S.K., Reed, S.C., Nemergut, D.R. et al. 2008. The earliest stages of ecosystem succession in high-elevation (5000 metres above sea level), recently deglaciated soils. *Proceedings of the Royal Society B: Biological Sciences* 275, 1653, 2793–2802. doi: 10.1098/rspb.2008.0808

Schumm, S.A. 1973. Geomorphic thresholds and complex response of drainage systems, in Morisawa, M. (ed.), *Fluvial Geomorphology*. George Allen & Unwin, London, UK, pp. 299–309.

Schumm, S.A. 1977. *The Fluvial System*. The Blackburn Press, New York.

Schumm, S.A. 1979. Geomorphic thresholds: The concept and its applications. *Transactions of the Institute of British Geographers* 4, 4, 485–515. doi: 10.2307/622211

Schumm, S.A. and Lichty, R.W. 1965. Time, space, and causality in geomorphology. *American Journal of Science* 263, 2, 110–119. doi: 10.2475/ajs.263.2.110

Schumm, S.A. and Parker, R.S. 1973. Implications of complex response of drainage systems for Quaternary alluvial stratigraphy. *Nature Physical Science* 243, 99–100. doi: 10.1038/physci243099a0

Scott, D.T., Baisden, W.T., Davies-Colley, R. et al. 2006. Localized erosion affects national carbon budget: Riverine carbon export from New Zealand. *Geophysical Research Letters* 33, 1, 1–4. doi: 10.1029/2005GL024644

Selva, E.C., Couto, E.G., Johnson, M.S. et al. 2007. Litterfall production and fluvial export in headwater catchments of the southern Amazon. *Journal of Tropical Ecology* 23, 3, 329–335. doi: 10.1017/S0266467406003956

Semiletov, I., Pipko, I., Gustafsson, Ö. et al. 2016. Acidification of East Siberian Arctic Shelf waters through addition of freshwater and terrestrial carbon. *Nature Geoscience* 9, 361–365. doi: 10.1038/ngeo2695

Semiletov, I.P., Pipko, I.I., Shakhova, N.E. et al. 2011. Carbon transport by the Lena River from its headwaters to the Arctic Ocean, with emphasis on fluvial input of terrestrial particulate organic carbon vs. carbon transport by coastal erosion. *Biogeosciences* 8, 2407–2426. doi: 10.5194/bg-8-2407-2011

Seo, J.I., Nakamura, F., Nakano, D. et al. 2008. Factors controlling the fluvial export of large woody debris, and its contribution to organic carbon budgets at watershed scales. *Water Resources Research* 44, 4, 1–13. doi: 10.1029/2007WR006453

Sharp, M., Parkes, J., Cragg, B. et al. 1999. Widespread bacterial populations at glacier beds and their relationship to rock weathering and carbon cycling. *Geology* 27, 2, 107–110. doi: 10.1130/0091-7613(1999)027<0107:WBPAGB>2.3.CO;2

Shi, Z., Allison, S.D., He, Y. et al.. 2020. The age distribution of global soil carbon inferred from radiocarbon measurements. *Nature Geoscience* 13, 555–559. doi: 10.1038/s41561-020-0596-z

Shields, F.D., Knight, S.S., Morin, N. et al. 2003. Response of fishes and aquatic habitats to sand-bed stream restoration using large woody debris. *Hydrobiologia* 494, 251–257. doi: 10.1007/978-94-017-3366-3_34

Short, F.T., Coles, R., Fortes, M.D. et al. 2014. Monitoring in the Western Pacific region shows evidence of seagrass decline in line with global trends. *Marine Pollution Bulletin* 83, 2, 408–416. doi: 10.1016/j.marpolbul.2014.03.036

Shuttleworth, E.L., Evans, M.G., Hutchinson, S.M. et al. 2015. Peatland restoration: Controls on sediment production and reductions in carbon and pollutant export. *Earth Surface Processes and Landforms* 40, 4, 459–472. doi: 10.1002/esp.3645

Sidle, R.C., Tsuboyama, Y., Noguchi, S. et al. 2000. Stormflow generation in steep forested headwaters: A linked hydrogeomorphic paradigm. *Hydrological Processes* 14, 3, 369–385. doi: 10.1002/(SICI)1099-1085(20000228)14:3<369::AID-HYP943>3.0.CO;2-P

Simon, C., Wiezer, A., Strittmatter, A.W. et al. 2009. Phylogenetic diversity and metabolic potential revealed in a glacier ice metagenome. *Applied and Environmental Microbiology* 75, 23, 7519–7526. doi: 10.1128/AEM.00946-09

Šímová, I. and Storch, D. 2017. The enigma of terrestrial primary productivity: Measurements, models, scales and the diversity–productivity relationship. *Ecography* 40, 2, 239–252. doi: 10.1111/ecog.02482

Singer, G.A., Fasching, C., Wilhelm, L. et al. 2012. Biogeochemically diverse organic matter in Alpine glaciers and its downstream fate. *Nature Geoscience* 5, 710–714. doi: 10.1038/ngeo1581

Singh, K.P., Mandal, T.N., and Tripathi, S.K. 2001. Patterns of restoration of soil physciochemical properties and microbial biomass in different landslide sites in the sal forest ecosystem of Nepal Himalaya. *Ecological Engineering* 17, 4, 385–401. doi: 10.1016/S0925-8574(00)00162-2

Sirvent, J., Desir, G., Gutierrez, M. et al. 1997. Erosion rates in badland areas recorded by collectors, erosion pins and profilometer techniques (Ebro Basin, NE-Spain). *Geomorphology* 18, 2, 61–75. doi: 10.1016/S0169-555X(96)00023-2

Skidmore, A.K. 1990. Terrain position as mapped from a gridded digital elevation model. *International Journal of Geographical Information Systems* 4, 1, 33–49. doi: 10.1080/02693799008941527

Skidmore, M., Anderson, S.P., Sharp, M. et al. 2005. Comparison of microbial community compositions of two subglacial environments reveals a possible role for microbes in chemical weathering processes. *Applied and Environmental Microbiology* 71, 11, 6986–6997. doi: 10.1128/AEM.71.11.6986-6997.2005

Skidmore, M.L., Foght, J.M., and Sharp, M.J. 2000. Microbial life beneath a high Arctic glacier. *Applied and Environmental Microbiology* 66, 8, 3214–3220. doi: 10.1128/AEM.66.8.3214-3220.2000

Skjemstad, J.O., Spouncer, L.R., Cowie, B. et al. 2004. Calibration of the Rothamsted organic carbon turnover model (RothC ver. 26.3), using measurable soil organic carbon pools. *Soil Research* 42, 1, 79–88. doi: 10.1071/SR03013

Slaymaker, O. 2003. The sediment budget as conceptual framework and management tool, in Kronvang, B. (ed.), *The Interactions Between Sediments and Water*. Springer, Dordrecht, The Netherlands, pp. 71–82.

Slaymaker, O. 2009. The future of geomorphology. *Geography Compass* 3, 1, 329–349. doi: 10.1111/j.1749-8198.2008.00178.x

Slaymaker, O. and Spencer, T. 1998. *Physical Geography and Global Environmental Change*. Longman, London, UK.

Slaymaker, O., Spencer, T., and Embleton-Hamann, C. 2009. *Geomorphology and Global Environmental Change*. Cambridge University Press, Cambridge, UK.

Smith, K.A., Ball, T., Conen, F. et al. 2003. Exchange of greenhouse gases between soil and atmosphere: Interactions of soil physical factors and biological processes. *European Journal of Soil Science* 54, 4, 779–791. doi: 10.1046/j.1351-0754.2003.0567.x

Smith, M.J. and Pain, C.F. 2009. Applications of remote sensing in geomorphology. *Progress in Physical Geography: Earth and Environment* 33, 4, 568–582. doi: 10.1177/0309133309346648

Smith, S.V., Renwick, W.H., Buddemeier, R.W. et al. 2001. Budgets of soil erosion and deposition for sediments and sedimentary organic carbon across the conterminous United States. *Global Biogeochemical Cycles* 15, 3, 697–707. doi: 10.1029/2000GB001341

Sollins, P., Homann, P., and Caldwell, B. 1996. Stabilization and destabilization of soil organic matter: Mechanisms and controls. *Geoderma* 74, 1–2, 65–105. doi: 10.1016/S0016-7061(96)00036-5

Song, C., Dannenberg, M.P., and Hwang, T. 2013. Optical remote sensing of terrestrial ecosystem primary productivity. *Progress in Physical Geography* 37, 6, 834–854. doi: 10.1177/0309133313507944

Song, Y., Deng, S.P., Acosta-Martínez, V. et al. 2008. Characterization of redox-related soil microbial communities along a river floodplain continuum by fatty acid methyl ester (FAME) and 16S rRNA genes. *Applied Soil Ecology* 40, 3, 499–509. doi: 10.1016/j. apsoil.2008.07.005

Spencer, R.G.M., Stubbins, A., Hernes, P.J. et al. 2009. Photochemical degradation of dissolved organic matter and dissolved lignin phenols from the Congo River. *Journal of Geophysical Research: Biogeosciences* 114, 1–12. doi: 10.1029/2009JG000968

Stallard, R.F. 1998. Terrestrial sedimentation and the carbon cycle: Coupling weathering and erosion to carbon burial. *Global Biogeochemical Cycles* 12, 2, 231–257. doi: 10.1029/98GB00741

Stallins, J.A. 2006. Geomorphology and ecology: Unifying themes for complex systems in biogeomorphology. *Geomorphology* 77, 3–4, 207–216. doi: 10.1016/j.geomorph.2006.01.005

Stanley, E.H., Powers, S.M., Lottig, N.R. et al. 2012. Contemporary changes in dissolved organic carbon (DOC) in human-dominated rivers: Is there a role for DOC management? *Freshwater Biology* 57, 26–42. doi: 10.1111/j.1365-2427.2011.02613.x

Steffen, W., Broadgate, W., Deutsch, L. et al. 2015. The trajectory of the Anthropocene: The Great Acceleration. *The Anthropocene Review* 2, 1, 81–98. doi: 10.1177/2053019614564785

Stevenson, F.J. 1994. *Humus Chemistry.* John Wiley & Sons, New York.

Stibal, M., Šabacká, M., and Žárský, J. 2012. Biological processes on glacier and ice sheet surfaces. *Nature Geoscience* 5, 771–774. doi: 10.1038/ngeo1611

Stibal, M., Tranter, M., Benning, L.G. et al. 2008. Microbial primary production on an Arctic glacier is insignificant in comparison with allochthonous organic carbon input. *Environmental Microbiology* 10, 8, 2172–2178. doi: 10.1111/j.1462-2920.2008.01620.x

Strahler, A.N. 1980. Systems theory in physical geography. *Physical Geography* 1, 1, 1–27. doi: 10.1080/02723646.1980.10642186

Sulman, B.N., Desai, A.R., Saliendra, N.Z. et al. 2010. CO_2 fluxes at northern fens and bogs have opposite responses to inter-annual fluctuations in water table. *Geophysical Research Letters* 37, 19, 1–5. doi: 10.1029/2010GL044018

Sundh, I., Nilsson, M., Granberg, G. et al. 1994. Depth distribution of microbial production and oxidation of methane in northern boreal peatlands. *Microbial Ecology* 27, 3, 253–265. doi: 10.1007/BF00182409

Sundquist, E.T. 1991. Steady-and non-steady-state carbonate-silicate controls on atmospheric CO_2. *Quaternary Science Reviews* 10, 2–3, 283–296. doi: 10.1016/0277-3791(91)90026-Q

Sutfin, N.A., Wohl, E.E., and Dwire, K.A. 2016. Banking carbon: A review of organic carbon storage and physical factors influencing retention in floodplains and riparian ecosystems. *Earth Surface Processes and Landforms* 41, 1, 38–60. doi: 10.1002/esp.3857

Swales, A., Bentley, S.J., and Lovelock, C.E. 2015. Mangrove-forest evolution in a sediment-rich estuarine system: Opportunists or agents of geomorphic change? *Earth Surface Processes and Landforms* 40, 12, 1672–1687. doi: 10.1002/esp.3759

Swanson, M.E., Franklin, J.F., Beschta, R.L. et al. 2010. The forgotten stage of forest succession: Early-successional ecosystems on forest sites. *Frontiers in Ecology and the Environment* 9, 2, 117–125. doi: 10.1890/090157

Swift, M.J., Heal, O.W., and Anderson, J.M. 1979. *Decomposition in Terrestrial Ecosystems.* University of California Press, Berkeley and Los Angeles, CA.

Swinnen, W., Daniëls, T., Maurer, E. et al. 2020. Geomorphic controls on floodplain sediment and soil organic carbon storage in a Scottish mountain river. *Earth Surface Processes and Landforms* 45, 1, 207–223. doi: 10.1002/esp.4729

Syvitski, J.P.M. and Kettner, A. 2011. Sediment flux and the Anthropocene. *Philosophical Transactions of the Royal Society A: Mathematical Physical and Engineering Sciences* 369, 1938, 957–975. doi: 10.1098/rsta.2010.0329

Takeuchi, N., Kohshima, S., and Seko, K. 2001. Structure, formation, and darkening process of albedo-reducing material (Cryoconite) on a Himalayan glacier: A granular algal mat growing on the glacier. *Arctic, Antarctic and Alpine Research* 33, 2, 115–122. doi: 10.2307/1552211

Tallis, J. 1997. Peat erosion in the Pennines: The badlands of Britain. *Biologist* 44, 277–279.

Tate, R.L. 1979. Effect of flooding on microbial activities in organic soils: Carbon metabolism. *Soil Science* 128, 5, 267–273. doi: 10.1097/00010694-197911000-00002

Telling, J., Anesio, A.M., Tranter, M. et al. 2012. Controls on the autochthonous production and respiration of organic matter in cryoconite holes on high Arctic glaciers: Carbon production on Arctic glaciers. *Journal of Geophysical Research: Biogeosciences* 117, G1, 1–10. doi: 10.1029/2011JG001828

Thatoi, H., Behera, B.C., and Mishra, R.R. 2013. Ecological role and biotechnological potential of mangrove fungi: A review. *Mycology* 4, 1, 54–71. doi: 10.1080/21501203.2013.785448

Theuerkauf, E.J., Stephens, J.D., Ridge, J.T. et al. 2015. Carbon export from fringing saltmarsh shoreline erosion overwhelms carbon storage across a critical width threshold. *Estuarine, Coastal and Shelf Science* 164, 367–378. doi: 10.1016/j.ecss.2015.08.001

Thomas, D.S.G. and Allison, R.J. (eds.). 1993. *Landscape Sensitivity.* John Wiley & Sons, Chichester, UK.

Thomazini, A., Teixeira, D.D.B., Turbay, C.V.G. et al. 2014. Spatial variability of CO_2 emissions from newly exposed paraglacial soils at a glacier retreat zone on King George Island, maritime Antarctica. *Permafrost and Periglacial Processes* 25, 4, 233–242. doi: 10.1002/ppp.1818

Thompson, J.A. and Kolka, R.K. 2005. Soil carbon storage estimation in a forested watershed using quantitative soil-landscape modeling. *Soil Society of America Journal* 69, 4, 1086–1093. doi: 10.2136/sssaj2004.0322

Thoms, M. 2003. Floodplain-river ecosystems: Lateral connections and the implications of human interference. *Geomorphology* 56, 3–4, 335–349. doi: 10.1016/S0169-555X(03)00160-0

Thornbush, M. 2015. Geography, urban geomorphology and sustainability. *Area* 47, 4, 350–353. doi: 10.1111/area.12218

Timpane-Padgham, B.L., Beechie, T., and Klinger, T. 2017. A systematic review of ecological attributes that confer resilience to climate change in environmental restoration. *PLOS ONE* 12, 3, 1–23. doi: 10.1371/journal.pone.0173812

Tockner, K., Pennetzdorfer, D., Reiner, N. et al. 1999. Hydrological connectivity, and the exchange of organic matter and nutrients in a dynamic river–floodplain system (Danube, Austria). *Freshwater Biology* 41, 3, 521–535. doi: 10.1046/j.1365-2427.1999.00399.x

Tockner, K., and Stanford, J.A. 2002. Riverine flood plains: Present state and future trends. *Environmental Conservation* 29, 3, 308–330. doi: 10.1017/S037689290200022X

Tockner, K., Ward, J.V., Arscott, D.B. et al. 2003. The Tagliamento River: A model ecosystem of European importance. *Aquatic Sciences* 65, 239–253. doi: 10.1007/s00027-003-0699-9

Torres, M., Limaye, A., Ganti, V. et al. 2017. Model predictions of long-lived storage of organic carbon in river deposits. *Earth Surface Dynamics Discussions* 5, 711–730. doi: 10.5194/esurf-5-711-2017

Torres, M.A., West, A.J., and Li, G. 2014. Sulphide oxidation and carbonate dissolution as a source of CO_2 over geological timescales. *Nature* 507, 346–349. doi: 10.1038/nature13030

Townend, I., Fletcher, C., Knappen, M. et al. 2011. A review of salt marsh dynamics. *Water and Environment Journal* 25, 4, 477–488. doi: 10.1111/j.1747-6593.2010.00243.x

Trammell, T.L.E. and Carreiro, M.M. 2012. Legacy effects of highway construction disturbance and vegetation management on carbon dynamics in forested urban verges, in Lal, R. and Augustin, B. (eds.), *Carbon Sequestration in Urban Ecosystems*. Springer, Dordrecht, Netherlands, pp. 331–352. doi: 10.1007/978-94-007-2366-5_17

Trevathan-Tackett, S.M., Seymour, J.R., Nielsen, D.A. et al. 2017. Sediment anoxia limits microbial-driven seagrass carbon remineralization under warming conditions. *FEMS Microbiology Ecology* 93, 6, 1–15. doi: 10.1093/femsec/fix033

Trimble, S.W. 1981. Changes in sediment storage in the Coon Creek Basin, driftless area, Wisconsin, 1853 to 1975. *Science* 214, 4517, 181–183. doi: 10.1126/science.214.4517.181

Trimble, S.W. 1983. A sediment budget for Coon Creek basin in the Driftless Area, Wisconsin, 1853–1977. *American Journal of Science* 283, 454–474. doi: 10.2475/ajs.283.5.454

Troxler, T. 2013. Integrated carbon budget models for the Everglades terrestrial-coastal-oceanic gradient: Current status and needs for inter-site comparisons. *Oceanography* 26, 3, 98–107. doi: 10.5670/oceanog.2013.51

Turner, M.G. 1989. Landscape ecology: The effect of pattern on process. *Annual Review of Ecology and Systematics* 20, 171–197. doi: 10.1146/annurev.es.20.110189.001131

Turowski, J.M., Hilton, R.G., and Sparkes, R. 2016. Decadal carbon discharge by a mountain stream is dominated by coarse organic matter. *Geology* 44, 1, 27–30. doi: 10.1130/G37192.1

Tweed, F.S., Russel, A.J., Warburton, J. et al. 2007. Research on sediment fluxes and sediment budgets in changing cold environments, in Beylich, A.A. and Warburton, J. (eds.), *Analysis of Source to Sink Fluxes and Sediment Budgets in Changing High Latitude and High Altitude Cold Environment*. Sediflux project, 158 p. https://www.ngu.no/FileArchive/237/2007_053.pdf

Unger, I.M., Kennedy, A.C., and Muzika, R.-M. 2009. Flooding effects on soil microbial communities. *Applied Soil Ecology* 42, 1, 1–8. doi: 10.1016/j.apsoil.2009.01.007

United Nations, Department of Economic and Social Affairs, Population Division 2014. World Urbanization Prospects: The 2014 revision, highlights (ST/ESA/SER.A/352).

Uuemaa, E., Mander, Ü., and Marja, R. 2013. Trends in the use of landscape spatial metrics as landscape indicators: A review. *Ecological Indicators* 28, 100–106. doi: 10.1016/j.ecolind.2012.07.018

Van Cappellen, P. and Maavara, T. 2016. Rivers in the Anthropocene: Global scale modifications of riverine nutrient fluxes by damming. *Ecohydrology & Hydrobiology* 16, 2, 106–111. doi: 10.1016/j.ecohyd.2016.04.001

Vance, D., Teagle, D.A.H., and Foster, G.L. 2009. Variable quaternary chemical weathering fluxes and imbalances in marine geochemical budgets. *Nature* 458, 493–496. doi: 10.1038/nature07828

Van Der Heijden, M.G.A., Bardgett, R.D., and Van Straalen, N.M. 2008. The unseen majority: Soil microbes as drivers of plant diversity and productivity in terrestrial ecosystems. *Ecology Letters* 11, 3, 296–310. doi: 10.1111/j.1461-0248.2007.01139.x

van der Wal, A. and de Boer, W. 2017. Dinner in the dark: Illuminating drivers of soil organic matter decomposition. *Soil Biology and Biochemistry* 105, 45–48. doi: 10.1016/j.soilbio.2016.11.006

Van Hemelryck, H., Govers, G., Van Oost, K. et al. 2011. Evaluating the impact of soil redistribution on the in situ mineralization of soil organic carbon. *Earth Surface Processes and Landforms* 36, 4, 427–438. doi: 10.1002/esp.2055

van Noordwijk, M., Cerri, C., Woomer, P.L. et al. 1997. Soil carbon dynamics in the humid tropical forest zone. *Geoderma* 79, 1–4, 187–225. doi: 10.1016/S0016-7061(97)00042-6

Vannote, R.L., Minshall, G.W., Cummins, K.W. et al. 1980. The river continuum concept. *Canadian Journal of Fisheries and Aquatic Science* 37, 1, 130–137. doi: 10.1139/f80-017

Van Oost, K., Quine, T.A., Govers, G. et al. 2007. The impact of agricultural soil erosion on the global carbon cycle. *Science* 318, 5850, 626–629. doi: 10.1126/science.1145724

Verburg, P.H., Erb, K.-H., Mertz, O. et al. 2013. Land system science: Between global challenges and local realities. *Current Opinion in Environmental Sustainability* 5, 5, 433–437. doi: 10.1016/j.cosust.2013.08.001

Vernberg, F.J. 1993. Salt-marsh processes: A review. *Environmental Toxicology and Chemistry* 12, 12, 2167–2195. doi: 10.1002/etc.5620121203

Vigiak, O., Borselli, L., Newham, L.T.H. et al. 2012. Comparison of conceptual landscape metrics to define hillslope-scale sediment delivery ratio. *Geomorphology* 138, 1, 74–88. doi: 10.1016/j.geomorph.2011.08.026

Vigier, N., Bourdon, B., Turner, S. et al.2001. Erosion timescales derived from U-decay series measurements in rivers. *Earth and Planetary Science Letters* 193, 3–4, 549–563. https://doi.org/10.1016/S0012-821X(01)00510-6

Viles, H.A. 1988. *Biogeomorphology*. Wiley Blackwell, Oxford, UK.

Viles, H.A. 2012. Microbial geomorphology: A neglected link between life and landscape. *Geomorphology* 157–158, 6–16. doi: 10.1016/j.geomorph.2011.03.021

Viles, H.A., Naylor, L.A., Carter, N.E.A. et al. 2008. Biogeomorphological disturbance regimes: Progress in linking ecological and geomorphological systems. *Earth Surface Processes and Landforms* 33, 9, 1419–1435. doi: 10.1002/esp.1717

Virkkala, A.-M., Virtanen, T., Lehtonen, A. et al. 2018. The current state of CO_2 flux chamber studies in the Arctic tundra: A review. *Progress in Physical Geography Earth and Environment* 42, 2, 162–184. doi: 10.1177/0309133317745784

Vonk, J.E., Sánchez-García, L., van Dongen, B.E. et al. 2012. Activation of old carbon by erosion of coastal and subsea permafrost in Arctic Siberia. *Nature* 489, 137–140. doi: 10.1038/nature11392

Wainright, S.C., Couch, C.A., and Meyer, J.L. 1992. Fluxes of bacteria and organic matter into a blackwater river from river sediments and floodplain soils. *Freshwater Biology* 28, 1, 37–48. doi: 10.1111/j.1365-2427.1992.tb00560.x

Wainwright, J. and Parsons, A.J. 2002. The effect of temporal variations in rainfall on scale dependency in runoff coefficients. *Water Resources Research* 38, 12, 1–7. doi: 10.1029/2000WR000188

Wainwright, J., Turnbull, L., Ibrahim, T.G. et al. 2011. Linking environmental régimes, space and time: Interpretations of structural and functional connectivity. *Geomorphology* 126, 3–4, 387–404. doi: 10.1016/j.geomorph.2010.07.027

Walker, J.C.G., Hays, P.B., and Kasting, J.F. 1981. A negative feedback mechanism for the long-term stabilization of earth's surface temperature. *Journal of Geophysical Research: Oceans* 86, C10, 9776–9782. doi: 10.1029/JC086iC10p09776

Walker, L.R. and Shiels, A.B. 2008. Post-disturbance erosion impacts carbon fluxes and plant succession on recent tropical landslides. *Plant and Soil* 313, 1–2, 205–216. doi: 10.1007/s11104-008-9692-3

Wallenstein, M.D. and Burns, R.G. 2011. Ecology of extracellular enzyme activities and organic matter degradation in soil: A complex community-driven process. *Methods of Soil Enzymology* 9, 35–55. doi: 10.2136/sssabookser9.c2

Walling, D.E. 1983. The sediment delivery problem. *Journal of Hydrology* 65, 1–3, 209–237. doi: 10.1016/0022-1694(83)90217-2

Walling, D.E. 2006. Human impact on land–ocean sediment transfer by the world's rivers. *Geomorphology* 79, 3–4, 192–216. doi: 10.1016/j.geomorph.2006.06.019

Walling, D.E. and Collins, A.L. 2008. The catchment sediment budget as a management tool. *Environmental Science and Policy* 11, 2, 136–143. doi: 10.1016/j.envsci.2007.10.004

Walling, D.E., Fang, D., Nicholas, A.P. et al. 2006. River flood plains as carbon sinks, in Rowan, J.S., Duck, R.W., and Werritty, A. (eds), *Sediment Dynamics and the Hydromorphology of Fluvial Systems*. International Association of Hydrological Sciences Publication 306. IAHS Press, Wallingford, UK, pp. 460–470.

Walling, D.E., Owens, P.N., and Leeks, G.J.L. 1998. The role of channel and floodplain storage in the suspended sediment budget of the River Ouse, Yorkshire, UK. *Geomorphology* 22, 3–4, 225–242. doi: 10.1016/S0169-555X(97)00086-X

Walling, D.E. and Quine, T.A. 1990. Calibration of caesium-137 measurements to provide quantitative erosion rate data. *Land Degradation & Development* 2, 3, 161–175. doi: 10.1002/ldr.3400020302

Walling, D.E. and Quine, T.A. 1993. Using Chernobyl-derived fallout radionuclides to investigate the role of downstream conveyance losses in the suspended sediment budget of the River Severn, United Kingdom. *Physical Geography* 14, 3, 239–253. doi: 10.1080/02723646.1993.10642478

Wang, M., Markert, B., and Shen, W. 2011. Microbial biomass carbon and enzyme activities of urban soils in Beijing. *Environmental Science and Pollution Research* 18, 958–967. doi: 10.1007/s11356-011-0445-0

Wang, Z., Govers, G., Steegen, A. et al. 2010. Catchment-scale carbon redistribution and delivery by water erosion in an intensively cultivated area. *Geomorphology* 124, 1–2, 65–74. doi: 10.1016/j.geomorph.2010.08.010

Warburton, J. 1990. An alpine proglacial fluvial sediment budget. *Geografisker Annaler Series A Physical Geography* 72, 3–4, 261–272. doi: 10.2307/521154

Washbourne, C.-L., Renforth, P., and Manning, D.A.C. 2012. Investigating carbonate formation in urban soils as a method for capture and storage of atmospheric carbon. *Science of the Total Environment* 431, 166–175. doi: 10.1016/j.scitotenv.2012.05.037

Wass, P.D. and Leeks, G.J.L. 1999. Suspended sediment fluxes in the Humber catchment, UK. *Hydrological Processes* 13, 7, 935–953. doi: 10.1002/(SICI)1099-1085(199905)13:7<935::AID-HYP783>3.0.CO;2-L

Waters, C.N., Zalasiewicz, J., and Summerhayes, C. 2016. The Anthropocene is functionally and stratigraphically distinct from the Holocene. *Science* 351, 6269, aad2622-1-aad2622-10. doi: 10.1126/science.aad2622

Waycott, M., Duarte, C.M., Carruthers, T.J.B. et al. 2009. Accelerating loss of sea-grasses across the globe threatens coastal ecosystems. *Proceedings of the National Academy of Sciences of the United States of America* 106, 30, 12377–12381. doi: 10.1073/pnas.0905620106

Welti, N., Bondar-Kunze, E., and Singer, G. 2012. Large-scale controls on potential respiration and denitrification in riverine floodplains. *Ecological Engineering* 42, 73–84. doi: 10.1016/j.ecoleng.2012.02.005

Weng, E., Luo, Y., Wang, W. et al. 2012. Ecosystem carbon storage capacity as affected by disturbance regimes: A general theoretical model. *Journal of Geophysical Research: Biogeosciences* 117, G3, 1–15. doi: 10.1029/2012JG02040

West, A.J. 2012. Thickness of the chemical weathering zone and implications for erosional and climatic drivers of weathering and for carbon-cycle feedbacks. *Geology* 40, 9, 811–814. doi: 10.1130/G33041.1

West, A.J., Bickle, M.J., Collins, R. et al. 2002. Small-catchment perspective on Himalayan weathering fluxes. *Geology* 30, 4, 355–358. doi: 10.1130/0091-7613(2002)030<0355:SCPOHW>2.0.CO;2

West, A.J., Galy, A., and Bickle, M. 2005. Tectonic and climatic controls on silicate weathering. *Earth and Planetary Science Letters* 235, 1–2, 211–228. doi: 10.1016/j.epsl.2005.03.020

West, T.O. and Post, W.M. 2002. Soil organic carbon sequestration rates by tillage and crop rotation. *Soil Society of America Journal* 66, 1930–1946. doi: 10.3334/CDIAC/tcm.002

Westbrook, C.J., Cooper, D.J., and Baker, B.W. 2011. Beaver assisted river valley formation. *River Research and Applications* 27, 2, 247–256. doi: 10.1002/rra.1359

Whiting, G.J. and Chanton, J.P. 2001. Greenhouse carbon balance of wetlands: Methane emission versus carbon sequestration. *Tellus B: Chemical and Physical Meteorology* 53, 5, 521–528. doi: 10.3402/tellusb.v53i5.16628

Wieder, W.R., Bonan, G.B., and Allison, S.D. 2013. Global soil carbon projections are improved by modelling microbial processes. *Nature Climate Change* 3, 909–912. doi: 10.1038/nclimate1951

Wiesmeier, M., Urbanski, L., Hobley, E. et al. 2019. Soil organic carbon storage as a key function of soils – A review of drivers and indicators at various scales. *Geoderma* 333, 149–162. doi: 10.1016/j.geoderma.2018.07.026

Willemsen, P.W.J.M., Horstman, E.M., Borsje, B.W. et al. 2016. Sensitivity of the sediment trapping capacity of an estuarine mangrove forest. *Geomorphology* 273, 189–201. doi: 10.1016/j.geomorph.2016.07.038

Willenbring, J.K. and Jerolmack, D.J. 2016. The null hypothesis: Globally steady rates of erosion, weathering fluxes and shelf sediment accumulation during late Cenozoic mountain uplift and glaciation. *Terra Nova* 28, 1, 11–18. doi: 10.1111/ter.12185

Wilson, J.S., Baldwin, D.S., Rees, G.N. et al. 2011. The effects of short-term inundation on carbon dynamics, microbial community structure and microbial activity in floodplain soil. *River Research and Applications* 27, 2, 213–225. doi: 10.1002/rra.1352

Winsborough, C. and Basiliko, N. 2010. Fungal and bacterial activity in northern peatlands. *Geomicrobiology Journal* 27, 4, 315–320. doi: 10.1080/01490450903424432

Wittmann, H. and von Blanckenburg, F. 2009. Cosmogenic nuclide budgeting of floodplain sediment transfer. *Geomorphology* 109, 3–4, 246–256. doi: 10.1016/j.geomorph.2009.03.006

Wohl, E. 2013a. Wilderness is dead: Whither critical zone studies and geomorphology in the Anthropocene? *Anthropocene* 2, 4–15. doi: 10.1016/j.ancene.2013.03.001

Wohl, E. 2013b. Landscape-scale carbon storage associated with beaver dams. *Geophysical Research Letters* 40, 14, 3631–3636. doi: 10.1002/grl.50710

Wohl, E., Dwire, K., Sutfin, N. et al. 2012. Mechanisms of carbon storage in mountainous headwater rivers. *Nature Communications* 3, 1–8. doi: 10.1038/ncomms2274

Wohl, E., Hall, R.O., Lininger, K.B. et al. 2017. Carbon dynamics of river corridors and the effects of human alterations. *Ecological Monographs* 87, 3, 379–409. doi: 10.1002/ecm.1261

Wohl, E., Lane, S.N., and Wilcox, A.C. 2015. The science and practice of river restoration. *Water Resources Research* 51, 8, 5974–5997. doi: 10.1002/2014WR016874

Wolman, M.G. and Leopold, L.B. 1957. River flood plains: Some observations on their formation. *United States Geological Society Professional Paper* 282C, 87–109. https://pubs.usgs.gov/pp/0282c/report.pdf

Wolman, M.G. and Miller, J.P. 1960. Magnitude and frequency of forces in geomorphic processes. *Journal of Geology* 68, 1, 54–74. doi: 10.1086/626637

Woodroffe, C. 1992. Mangrove sediments and geomorphology, in Robertson, A.I., and Alongi, D.M. (eds.), *Tropical Mangrove Ecosystems*. American Geophysical Union, Washington, DC, pp. 7–41.

Woodroffe, C.D., Rogers, K., McKee, K.L. et al. 2016. Mangrove sedimentation and response to relative sea-level rise. *Annual Review of Marine Science* 8, 243–266. doi: 10.1146/annurev-marine-122414-034025

World Bank 2017. FAO electronic files and website. World Development Indicators http://databank.worldbank.org/data/home.aspx (accessed August 2017).

Worrall, F. and Burt, T.P. 2007. Flux of dissolved organic carbon from U.K. rivers. *Global Biogeochemical Cycles* 21, 1, 1–14. doi: 10.1029/2006GB002709

Worrall, F., Burt, T.P., Howden, N.J.K. et al. 2014. Variation in suspended sediment yield across the UK – A failure of the concept and interpretation of the sediment delivery ratio. *Journal of Hydrology* 519, 1985–1996. doi: 10.1016/j.jhydrol.2014.09.066

Worrall, F., Burt, T.P., Rowson, J.G. et al. 2009. The multi-annual carbon budget of a peat-covered catchment. *Science of the Total Environment* 407, 13, 4084–4094. doi: 10.1016/j.scitotenv.2009.03.008

Worrall, F., Davies, H., Bhogal, A. et al. 2012. The flux of DOC from the UK – Predicting the role of soils, land use and net watershed losses. *Journal of Hydrology* 448–449, 149–160. doi: 10.1016/j.jhydrol.2012.04.053

Worrall, F., Reed, M., Warburton, J. et al. 2003. Carbon budget for a British upland peat catchment. *Science of the Total Environment* 312, 1–3, 133–146. doi: 10.1016/S0048-9697(03)00226-2

Wu, X. and Murray, A.T. 2008. A new approach to quantifying spatial contiguity using graph theory and spatial interaction. *International Journal of Geographical Information Science* 22, 4, 387–407. doi: 10.1080/13658810701405615

Xiang, S.-R., Shang, T.-C., Chen, Y. et al. 2009. Deposition and postdeposition mechanisms as possible drivers of microbial population variability in glacier ice. *FEMS Microbiology Ecology* 70, 2, 165–176. doi: 10.1111/j.1574-6941.2009.00759.x

Xu, J., Morris, P.J., Liu, J. et al. 2018. PEATMAP: Refining estimates of global peatland distribution based on a meta-analysis. *Catena* 160, 134–140. doi: 10.1016/j.catena.2017.09.010

Yuste, J.C., Peñuelas, J., Estiarte, M. et al. 2011. Drought-resistant fungi control soil organic matter decomposition and its response to temperature. *Global Change Biology* 17, 3, 1475–1486. doi: 10.1111/j.1365-2486.2010.02300.x

Zalasiewicz, J., Waters, C.N., Summerhayes, C.P. et al. 2017. The Working Group on the Anthropocene: Summary of evidence and interim recommendations. *Anthropocene* 19, 55–60. doi: 10.1016/j.ancene.2017.09.001

Zalasiewicz, J., Williams, M., Smith, A. et al. 2008. Are we now living in the Anthropocene? *Geological Society of America Today* 18, 2, 4–8. doi: 10.1130/GSAT01802A.1

Zeebe, R.E. and Caldeira, K. 2008. Close mass balance of long-term carbon fluxes from ice-core CO_2 and ocean chemistry records. *Nature Geoscience* 1, 312–315. doi: 10.1038/ngeo185

Zehetner, F., Lair, G.J., and Gerzabek, M.H. 2009. Rapid carbon accretion and organic matter pool stabilization in riverine floodplain soils. *Global Biogeochemical Cycles* 23, 4, 1–7. doi: 10.1029/200GB003481

Zhang, X., Schumann, M., Gao, Y. et al. 2016. Restoration of high-altitude peatlands on the Ruoergai Plateau, Northeastern Tibetan Plateau, China, in Bonn, A., Allott, T.E.H., and Evans, M. et al. (eds.), *Peatland Restoration and Ecosystem Services: Science Policy and Practice*. Cambridge University Press, Cambridge, UK, pp. 234–252.

Zhang, X., Wu, S., Cao, W. et al. 2015. Dependence of the sediment delivery ratio on scale and its fractal characteristics. *International Journal of Sediment Research* 30, 4, 338–343. doi: 10.1016/j.ijsrc.2015.03.011

Zigah, P.K., McNichol, A.P., Xu, L. et al. 2017. Allochthonous sources and dynamic cycling of ocean dissolved organic carbon revealed by carbon isotopes: Carbon isotopes of marine DOC. *Geophysical Research Letters* 44, 5, 2407–2415. doi: 10.1002/2016GL071348

Zonneveld, I.S. 1989. The land unit – A fundamental concept in landscape ecology, and its applications. *Landscape Ecology* 3, 2, 67–86. doi: 10.1007/BF00131171

Zumsteg, A., Luster, J., Göransson, H. et al. 2012. Bacterial, archaeal and fungal succession in the forefield of a receding glacier. *Microbial Ecology* 63, 3, 552–564. doi: 10.1007/s00248-011-9991-8

Index

Geomorphology and the Carbon Cycle, First Edition. Martin Evans.
© 2022 Royal Geographical Society (with the Institute of British Geographers). Published 2022 by John Wiley & Sons Ltd.